Unleashing the Beast

A Comprehensive History of the Barracuda

Todd Bandel

Copyright © 2024 Todd Bandel

All rights reserved.

ISBN: 9798300664817

DEDICATION

I dedicate this book to all my past automotive mentors and colleagues. Your guidance, support, and shared wisdom have been invaluable in shaping my journey. Each of you played a significant role in my professional development, imparting knowledge and fostering a passion for excellence in the automotive field.

Content

ACKNOWLEDGEMENTS   I

CHAPTER ONE
The Birth of the Barracuda: Origins and Inception   1

Chapter Two
First Generation (1964-1966): A Fastback Revolution   13

Chapter Three
Second Generation (1967-1969): Growing Muscles   25

Chapter Four
Third Generation (1970-1974): The Peak of Pony Car Power   41

Chapter Five
Under the Hood: Engines that Defined the Barracuda   55

Chapter Six
Design Evolution: From Valiant-based to Standalone Stunner   69

Chapter Seven
Racing Pedigree: The Barracuda on Track and Strip   83

Chapter Eight
Rare and Coveted: Limited Editions and Special Models   99

Chapter Nine
The Barracuda's Place in the Muscle Car Era   113

Chapter Ten
Competitors and Contemporaries: How the 'Cuda Stacked Up   127

Chapter Eleven
The Barracuda's Cultural Impact and Legacy   141

Chapter Twelve
Collecting and Restoring: The Barracuda Today   157

ACKNOWLEDGMENTS

I want to express my deepest gratitude to my father for introducing me to the exhilarating world of automotive racing. Your passion for cars and dedication to the sport have inspired me.

From the first time you took me to a race track, I was captivated by the power and precision of the machines, as well as the skill required to master them.

Your guidance and support have fueled my interest and enthusiasm, making every moment in this thrilling world more meaningful. Thank you for sharing this incredible journey with me and for being such a pivotal influence in my life.

Unleashing the Beast: A Comprehensive History of the Barracuda

Chapter 1: The Birth of the Barracuda: Origins and Inception

Section 1.1: The Compact Car Craze

The early 1960s marked a significant shift in the American automotive landscape. As the decade dawned, a new trend was sweeping across the nation: the compact car craze. This movement represented a departure from the large, gas-guzzling behemoths that had dominated American roads in the previous decade.

At the forefront of this revolution was the Ford Falcon, introduced in 1960. The Falcon's success was nothing short of phenomenal, quickly becoming America's best-selling compact car. Its popularity sent shockwaves through the industry, forcing other manufacturers to take notice and respond. Chrysler, not content to sit on the sidelines, answered with the introduction of the Valiant in the same year.

The Valiant, while a solid entry into the compact car market, lacked the sporty appeal that was beginning to captivate younger buyers. As the baby boomer generation came of age, there was a growing demand for vehicles that offered more than just practicality.

They wanted cars that were fun to drive, stylish, and could make a statement.

This shift in consumer preferences didn't go unnoticed by automotive executives and designers. Market research began to reveal a clear trend: young buyers were seeking cars that combined the economy and maneuverability of compacts with the performance and style of sportier models. The stage was set for a new category of vehicle that would bridge this gap.

Plymouth, as a division of Chrysler, recognized the need to evolve beyond the practical but somewhat staid Valiant. They needed a car that could capture the imagination of younger buyers while still maintaining the reliability and affordability that had made the Valiant successful.

As the compact car craze continued to gain momentum, it became clear that the next big opportunity lay in creating a sporty compact car that could offer the best of both worlds. This realization would set the stage for one of the most exciting periods in American automotive history and ultimately lead to the birth of the Plymouth Barracuda.

The compact car craze wasn't just a fad; it was a fundamental shift in the American automotive market. It represented a change in consumer values, a response to economic pressures, and a reflection of the changing demographics of car buyers. For Plymouth, and indeed for all of Detroit, the challenge was clear: adapt to this new reality or risk being left behind.

Section 1.2: The Ford Mustang Threat

In the early 1960s, a seismic shift was underway in the American automotive landscape, and at its epicenter was Ford Motor Company. The Blue Oval had been secretly developing a revolutionary new vehicle that would soon reshape the entire industry: the Ford Mustang.

Unleashing the Beast: A Comprehensive History of the Barracuda

Word of Ford's upcoming pony car quickly spread through Detroit's automotive circles, reaching the ears of Chrysler executives. The intelligence gathered on the Mustang sent shockwaves through Plymouth's headquarters. It was clear that this new Ford offering had the potential to capture the hearts and wallets of young American car buyers, a demographic that every automaker coveted.

As details of the Mustang's development emerged, Chrysler recognized the urgent need to respond. The race was on to beat Ford to market with a comparable sporty, compact car. Plymouth, Chrysler's division aimed at younger buyers, was tasked with leading this charge. The pressure to innovate quickly was immense, as every day counted in the fast-paced world of automotive development.

Plymouth's strategy to compete with the impending Mustang threat was multi-faceted. First and foremost, they needed to leverage their existing resources efficiently. This meant utilizing the Valiant platform as a starting point, a decision that would prove both beneficial and limiting. The Valiant's established production lines and proven mechanicals would allow for a quicker development process, but it also meant working within certain constraints.

Secondly, Plymouth knew it needed to offer something unique to stand out from the Mustang. This led to the emphasis on the fastback design, which would become the Barracuda's signature feature. The design team was pushed to their creative limits, tasked with creating a car that was both visually striking and functionally superior to the anticipated Mustang.

The pressure to innovate quickly cannot be overstated. Traditional automotive development cycles were compressed, with designers and engineers working around the clock to bring the Barracuda from concept to reality in record time. This accelerated timeline meant making crucial decisions rapidly, sometimes without the luxury of extensive testing or market research.

Despite the challenges, the impending Mustang threat also created a sense of excitement and purpose within Plymouth. It was a David versus Goliath scenario, with the scrappy Plymouth team determined to prove they could go toe-to-toe with the industry giant Ford. This underdog mentality fueled creativity and fostered a spirit of innovation that would come to define the Barracuda project.

As the Mustang's launch drew nearer, the intensity of Plymouth's efforts increased. Every aspect of the car was scrutinized and refined, from the powertrain options to the interior trim. The goal was not just to compete with the Mustang, but to offer a compelling alternative that could carve out its own niche in the burgeoning pony car market.

Ultimately, the Ford Mustang threat catalyzed one of the most exciting periods in Plymouth's history. It pushed the company to new heights of innovation and daring, resulting in a car that would leave an indelible mark on automotive history. The Barracuda may have been born out of a need to compete, but it would go on to become a legend in its own right, beloved by enthusiasts for generations to come.

Section 1.3: Conception of the Barracuda

The conception of the Plymouth Barracuda was a pivotal moment in American automotive history, born out of necessity and innovation. As Ford's Mustang loomed on the horizon, Plymouth's design team knew they needed something truly special to compete in the burgeoning pony car market.

Initial design proposals for the Barracuda were varied and ambitious. Some designers advocated for a completely new platform, while others suggested more modest modifications to existing models. After much deliberation, a critical decision was made to base the new car on the Valiant platform. Both practical and financial considerations drove this choice, as it would allow for a quicker development process and lower production costs.

Unleashing the Beast: A Comprehensive History of the Barracuda

The key players in Barracuda's development were a mix of seasoned automotive veterans and fresh talent. Elwood Engel, Chrysler's chief stylist at the time, oversaw the project, while young designer Irv Ritchie was instrumental in crafting the car's distinctive silhouette. Their combined expertise and creativity would prove crucial in bringing the Barracuda to life.

The design process was not without its challenges. The team faced the daunting task of transforming the rather pedestrian Valiant into a sleek, sporty coupe that could capture the public's imagination. One of the biggest hurdles was incorporating a large, wraparound rear window, a feature that would become the Barracuda's signature design element. This innovative glass panel required new manufacturing techniques and posed significant engineering challenges.

The most important aspect of the Barracuda's conception was the emphasis placed on its distinctive fastback design. This sleek profile not only set the car apart visually from its Valiant sibling but also gave it a sporty, aggressive stance that would appeal to younger buyers. The fastback design was a bold statement, one that would define the Barracuda and influence pony car aesthetics for years to come.

As the design process progressed, the team worked tirelessly to refine every aspect of the car. From the curve of the fenders to the placement of the headlights, each element was scrutinized and optimized. The interior received equal attention, with a focus on creating a driver-centric cockpit that would enhance the car's sporty feel.

The conception of the Barracuda was a testament to the creativity and determination of Plymouth's design and engineering teams. In a relatively short time, they had taken a spark of an idea and transformed it into a fully realized automobile that was poised to make its mark on the automotive world. As the final designs were approved and preparations for production began, a palpable sense of excitement prevailed. The Plymouth Barracuda was no longer just a concept; it was about to become a reality.

Section 1.4: Engineering Innovations

The Plymouth Barracuda wasn't just a pretty face; it was a testament to innovative engineering that set it apart from its competitors. At the heart of its distinctive design was the unique wraparound rear glass, a feature that would become the Barracuda's signature element.

This expansive piece of curved glass was not only visually striking but also presented significant engineering challenges. The team had to work closely with glass manufacturers to develop a production process that could consistently create this complex shape while maintaining structural integrity and optical clarity.

Adapting the Valiant's chassis and drivetrain for the Barracuda was another crucial engineering feat. The team had to strike a delicate balance between utilizing existing components to keep costs down and modifying them sufficiently to create a sportier driving experience. This involved tweaking suspension geometry, adjusting spring rates, and fine-tuning the steering system to deliver more responsive handling without sacrificing ride comfort.

Engine options were a key consideration in the development of the Barracuda. The engineering team knew that to compete in the emerging pony car market, they needed to offer a range of power plants to suit different customer preferences. The base model featured the reliable but modest Slant-6 engine, while performance enthusiasts could opt for more powerful V8 options. Each engine choice required careful integration into the Valiant-derived chassis, with considerations for cooling, exhaust routing, and weight distribution.

The interior design of the Barracuda was another area where engineering innovation shone. The team faced the challenge of creating a sporty, driver-focused cockpit within the constraints of the Valiant's basic architecture. They developed a new instrument panel with a sportier layout and incorporated bucket seats as standard, a feature that was still considered premium in many competitors. The folding rear seat was a particularly clever touch, allowing the Barracuda to offer practical cargo space when needed, a feature that set it apart from many sports cars of the era.

The most significant engineering challenge was balancing the Barracuda's sporting pretensions with practical considerations. The team had to ensure that the car delivered on its promise of performance and style without sacrificing the reliability and everyday usability that Plymouth customers expected. This required extensive testing and refinement of components, such as the braking system, which needed to be robust enough to handle spirited driving while still providing smooth, predictable performance in daily use.

The engineering innovations that went into the first Barracuda laid the groundwork for its future evolution. Many of the solutions developed for this first generation, from the unique rear glass to the adaptable chassis, would continue to influence Plymouth's design and engineering philosophy for years to come. These innovations not only made the Barracuda a compelling product at launch but also provided a solid foundation for the model to grow into a true muscle car icon in subsequent generations.

Section 1.5: Naming the Beast

The process of naming a new car model is a crucial step in its development, and the Plymouth Barracuda was no exception. As the innovative fastback neared completion, Plymouth executives and marketing teams faced the challenge of finding the perfect name that would capture the essence of their new creation.

The search for a name began with brainstorming sessions that produced dozens of potential monikers. The team sought a name that would evoke power, speed, and unique qualities that embodied their new sporty offering. After much deliberation, they settled on "Barracuda," a name that perfectly encapsulated the car's sleek, aggressive nature.

The barracuda fish, known for its fierce predatory behavior and streamlined body, served as an apt metaphor for Plymouth's new fastback. The name conjured images of a swift, powerful creature cutting through the water or, in this case, the streets. It was a name that promised performance and stood out in a sea of automotive appellations. However, "Barracuda" wasn't the only name on the table. Several alternatives were seriously considered during the naming process.

"Panda" was one such option, likely chosen for its alliterative quality with Plymouth, but it lacked the aggressive connotation the team desired. "Cheetah" was another contender, evoking speed and agility, but it was ultimately discarded, possibly due to its similarity to other feline-inspired car names of the era.

Marketing considerations played a significant role in the final decision. The name needed to appeal to the target demographic of young, performance-minded buyers. It had to be memorable, easy to pronounce, and lend itself well to advertising campaigns. "Barracuda" ticked all these boxes, offering a wealth of potential for creative marketing strategies.

When the name was finally unveiled to the public, the reception was largely positive. Car enthusiasts appreciated the aggressive connotations, and the general public found it intriguing and distinctive. The name quickly caught on, with fans and automotive journalists alike often shortening it to "Cuda," a nickname that would gain even more popularity in later years.

The choice of "Barracuda" proved to be a masterstroke for Plymouth. It set the car apart from its competitors and created an instant identity for the new model. The name became synonymous with the unique fastback design and performance aspirations of Plymouth's latest offering, helping to establish the Barracuda as a formidable presence in the burgeoning pony car market. In the years that followed, the Barracuda name would become an integral part of muscle car lore, representing not just a car, but a legacy of performance and style. The careful consideration given to naming this new model paid dividends, creating a brand identity that would endure long after the last Barracuda rolled off the production line.

Section 1.6: Pre-Launch Preparations

As the Plymouth Barracuda neared its highly anticipated debut, the Chrysler Corporation kicked into high gear with its pre-launch preparations. The final months leading up to the release were a whirlwind of activity, with every aspect of the car's introduction meticulously planned and executed. First and foremost, the engineering team conducted extensive final testing and refinements.

Every component of the Barracuda was scrutinized and fine-tuned to ensure it met the high standards set by Plymouth. From the distinctive wraparound rear glass to the adapted Valiant chassis, no detail was too small to escape attention. The team worked tirelessly to address any last-minute issues, making subtle adjustments to enhance performance, reliability, and overall quality.

Simultaneously, Plymouth's manufacturing division was busy setting up the production. The integration of the Barracuda into existing facilities required careful planning and execution. New tooling was installed, assembly processes were optimized, and workers were trained to handle the unique aspects of the latest model. The goal was to ensure a smooth transition from prototype to mass production, maintaining consistent quality across every unit that rolled off the line.

The dealer network played a crucial role in the Barracuda's launch, and Plymouth invested significant resources in preparing them for the new model. Comprehensive training programs were developed to educate sales staff about Barracuda's features, target market, and competitive advantages. Dealer showrooms were updated with new signage and promotional materials to create buzz around the upcoming release. Plymouth also worked closely with dealerships to establish optimal inventory levels and pricing strategies, maximizing the launch's impact.

Marketing and advertising strategies were carefully crafted to position the Barracuda as a desirable, sporty alternative in the compact car segment. Plymouth's advertising agency developed a multi-faceted campaign that included print ads, television commercials, and radio spots. The messaging emphasized the Barracuda's unique fastback design, versatility, and performance capabilities. Teasers were strategically released to build anticipation and generate word-of-mouth excitement among potential buyers.

In the weeks leading up to the official unveiling, Plymouth organized a series of press previews to give automotive journalists an exclusive first look at the Barracuda. These events were carefully orchestrated to showcase the car's best features and generate favorable coverage. Journalists were allowed to examine the car up close, speak with the designers and engineers, and in some cases, even take it for a short drive. The initial reactions were closely monitored and analyzed, with Plymouth using this feedback to fine-tune its launch strategy and address any concerns raised by the media.

These pre-launch preparations were critical in setting the stage for the Barracuda's grand entrance into the automotive market. By meticulously attending to every detail from engineering refinements to dealer preparation to media relations, Plymouth aimed to give its new creation the best possible start in life.

Unleashing the Beast: A Comprehensive History of the Barracuda

The company knew that first impressions would be crucial in establishing the Barracuda as a serious contender in the increasingly competitive world of American muscle cars. With the groundwork laid and anticipation at a fever pitch, all that remained was for Plymouth to pull back the curtain and introduce the world to the Barracuda.

Section 1.7: The Grand Unveiling

The Plymouth Barracuda made its grand debut on April 1, 1964, just two weeks before the Ford Mustang's launch. This strategic timing was no coincidence, as Plymouth aimed to establish itself in the sporty compact car market before its main competitor could steal the spotlight. The official unveiling was a carefully orchestrated event that took place simultaneously across Plymouth dealerships nationwide, creating a buzz of excitement among car enthusiasts and the general public alike.

As the covers were pulled off the gleaming new Barracudas in showrooms across America, the reaction was immediate and electric. The car's distinctive fastback design, dominated by the expansive wraparound rear window, drew gasps of admiration from onlookers. Many were struck by the Barracuda's sleek lines and sporty stance, which set it apart from the more conservative Valiant on which it was based.

Media representatives scrambled to get their first look at Plymouth's new offering, and initial reviews were largely positive. Automotive journalists praised the Barracuda's innovative design, particularly the unique rear glass that offered both style and practicality. The car's performance capabilities, while not revolutionary, were deemed competitive for its class, with reviewers noting its nimble handling and responsive engine options.

Inevitably, comparisons to the Ford Mustang began almost immediately. While the Mustang would ultimately prove to be the more popular car, many critics at the time gave the Barracuda high marks for its distinctive styling and versatile interior space. Some even preferred Plymouth's more subdued approach to sportiness, viewing it as a more mature and refined option in the segment.

Initial sales figures for the Barracuda were encouraging, if not spectacular. In its first year, Plymouth sold approximately 23,000 units, a respectable figure for a new model, though the Mustang's runaway success would overshadow it. However, these early sales numbers were enough to convince Plymouth executives that they had made the right decision in bringing the Barracuda to market.

The introduction of the Barracuda had a significant impact on Plymouth's image and market position. It gave the brand a much-needed injection of youth and sportiness, helping to shed its reputation for producing primarily practical, family-oriented vehicles. The Barracuda became the halo car for Plymouth, attracting younger buyers into showrooms and generating interest in the brand's other offerings.

Within the Plymouth lineup, the Barracuda occupied a unique position. It was sportier than the Valiant but more practical than an actual sports car, offering a blend of performance, style, and utility that appealed to a broad range of customers. This positioning proved crucial in the coming years, as it enabled Plymouth to refine and develop the Barracuda concept to meet changing market demands.

The grand unveiling of the Plymouth Barracuda marked a pivotal moment not just for Plymouth but for the American automotive industry as a whole. It signaled the beginning of the pony car era, a period of intense competition and innovation that would produce some of the most iconic vehicles in automotive history. The Barracuda's debut set the stage for its evolution into a true muscle car legend, a journey that had only just begun on that spring day in 1964.

Chapter 2: First Generation (1964-1966): A Fastback Revolution

Section 2.1: The Birth of the Fastback

The mid-1960s marked a pivotal moment in American automotive design, and at the forefront of this revolution was the fastback. The Plymouth Barracuda, introduced in 1964, was one of the pioneers of this bold new style. But what exactly is a fastback, and why was it so revolutionary?

A fastback is characterized by a single, smooth line that runs from the roof to the rear of the car, creating a sleek, aerodynamic profile. This design was a radical departure from the boxy sedans and wagons that dominated the roads at the time. The fastback offered a sporty, youthful appearance that resonated with a new generation of car buyers who were looking for something different.

Plymouth's decision to enter the sporty car market was not made in isolation. The company had been closely monitoring market trends and conducting extensive research. They recognized a growing demand for cars that offered both style and performance, particularly among younger buyers. This insight was further fueled by intelligence about Ford's development of the Mustang, which was set to create an entirely new market segment.

In a stroke of brilliance, Plymouth decided to base its new sporty offering on the existing Valiant platform. This decision allowed for a quicker development process and helped keep costs down. The Barracuda shared many components with the Valiant, including the basic unitized body structure, suspension, and drivetrain. However, the similarities ended there, as the Barracuda's distinctive fastback styling set it apart not just from the Valiant, but from every other car on the road.

The first-generation Barracuda featured several distinctive design elements that set it apart. Most notable was its massive rear window, which at the time was the largest ever used on a production car. This expansive glass area not only contributed to the car's sleek profile but also provided excellent visibility, creating a bright and airy interior. The Barracuda also featured a distinctive grille design and elegant side trim that further differentiated it from its Valiant sibling.

When the Barracuda hit showrooms in April 1964, just two weeks before the Ford Mustang, it created quite a stir. The public and critics alike were intrigued by its innovative design. Automotive journalists praised its unique styling and versatile interior, with the fold-down rear seat receiving particular acclaim for its practicality.

However, the Barracuda's reception was somewhat overshadowed by the launch of the Ford Mustang. While the Barracuda's sales were respectable, with 23,443 units sold in its first year, they paled in comparison to the Mustang's runaway success of over 400,000 units. Despite this, the Barracuda had succeeded in establishing Plymouth as a player in the nascent pony car market.

The birth of the Barracuda marked the beginning of a new era for Plymouth. It demonstrated the company's ability to innovate and respond to changing market demands. While it may not have achieved the same level of commercial success as some of its competitors, the first-generation Barracuda laid the groundwork for what would become one of the most iconic muscle cars of all time. Its influence on automotive design and its role in shaping the pony car segment cannot be overstated.

Section 2.2: 1964 Model Year Specifics

The inaugural year of the Plymouth Barracuda set the stage for what would become an iconic American muscle car. The 1964 model year offered a range of options that catered to various driving preferences and performance needs, establishing the Barracuda as a versatile contender in the burgeoning pony car market.

At the heart of the 1964 Barracuda was a selection of engine options that balanced economy with performance. The base powerplant was the reliable 170 cubic inch (2.8 L) Slant-6 engine, a stalwart of Chrysler's lineup known for its durability and efficiency.

This engine was perfect for drivers who prioritized fuel economy and low maintenance costs. However, for those seeking more power, Plymouth offered an optional 273 cubic inch (4.5 L) V8 engine. This V8 significantly boosted the Barracuda's performance, transforming it from a stylish commuter to a respectable performer capable of holding its own against other sporty cars of the era.

The transmission choices for the 1964 Barracuda further enhanced its appeal to a wide range of drivers. The standard offering was a three-speed manual transmission, which provided an engaging driving experience for those who preferred to shift gears themselves. For even more control and performance, Plymouth offered an optional four-speed manual transmission.

This option was particularly popular among driving enthusiasts who wanted to extract maximum performance from their Barracudas. For those who preferred a more relaxed driving experience, a three-speed automatic transmission was also available. This automatic

Unleashing the Beast: A Comprehensive History of the Barracuda

option made the Barracuda an excellent choice for daily driving and long-distance cruising, offering smooth shifts and effortless operation.

The interior of the 1964 Barracuda was a blend of sportiness and practicality. One of the most notable features was the fold-down rear seat, which was a revolutionary concept at the time. This innovative design allowed the Barracuda to transform from a comfortable four-seater into a cargo-hauling machine.

With the rear seat folded down, the Barracuda could accommodate long items through its large rear glass, making it surprisingly versatile for a sporty car. The interior also featured a modern dashboard design with easy-to-read gauges and controls within easy reach of the driver. Exterior styling cues of the 1964 Barracuda were distinctive and eye-catching.

The front end featured a unique grille design that set it apart from its Valiant sibling and other competitors. The side profile was adorned with sleek trim that accentuated the car's fastback silhouette. But the most striking feature was undoubtedly the massive rear window.

This expansive piece of glass was the largest ever used on a production car at the time, covering nearly the entire rear section of the roof. It not only provided excellent visibility but also became Barracuda's signature design element, instantly recognizable and widely admired.

Performance specifications for the 1964 Barracuda varied depending on the chosen configuration. With the base Slant-6 engine, the Barracuda was more about style than speed, boasting 0-60 mph times in the mid-14-second range.

However, when equipped with the 273 V8, particularly with the four-speed manual transmission, the Barracuda could sprint from 0-60 mph in around 10 seconds and complete the quarter-mile in approximately 17 seconds. While these figures may seem modest by today's standards, they were respectable for an affordable sports coupe in 1964.

The 1964 Plymouth Barracuda represented a bold entry into the sporty car market. Its combination of distinctive styling, versatile interior, and range of powertrain options laid the foundation for future success. While the Ford Mustang may have overshadowed it in terms of sales volume, the 1964 Barracuda established itself as a unique and compelling alternative in the pony car segment, setting the stage for the evolution of one of America's most beloved muscle cars.

Section 2.3: 1965 Model Year Updates

The 1965 model year brought significant improvements to the Plymouth Barracuda, cementing its position in the rapidly evolving pony car market. As the automotive landscape continued to shift, Plymouth recognized the need to refine and enhance its offering to stay competitive.

One of the most notable updates for 1965 was the introduction of the new Commando 273 engine option. This high-performance variant of the 273 cubic inch V8 featured a four-barrel carburetor, higher compression ratio, and a more aggressive camshaft profile. The result was a substantial boost in power, with output increasing from 180 to 235 horsepower. This engine option provided performance enthusiasts with a more potent powerplant, narrowing the gap between the Barracuda and some of its more powerful competitors.

Exterior styling received subtle yet effective tweaks for the 1965 model year. The front grille was revised with a more pronounced horizontal bar design, giving the car a broader and more aggressive appearance. New color options were introduced, including vibrant hues such as Rally Red and Electric Blue, allowing buyers to personalize their Barracudas further. The side trim was also modified, featuring a more pronounced spear-like design that accentuated the car's sleek profile.

Unleashing the Beast: A Comprehensive History of the Barracuda

The most significant addition to the 1965 Barracuda lineup was the introduction of the Formula S package. This performance-oriented option transformed the Barracuda from a stylish fastback into a true contender in the performance car segment. The Formula S package included the aforementioned Commando 273 engine, along with various suspension upgrades.

Stiffer springs, heavy-duty shock absorbers, and a larger front sway bar significantly improved the car's handling characteristics. The package also featured wider 14-inch wheels with high-performance tires, providing better grip and cornering ability. To complete the performance-oriented look, Formula S cars received special badging and racing stripes.

Despite these improvements, the Barracuda faced stiff competition in the sales department. The Ford Mustang, launched just after the Barracuda in 1964, had become a runaway success. While Barracuda sales did see a modest increase in 1965, they were still dwarfed by the Mustang's figures. Plymouth sold approximately 64,596 Barracudas in 1965, compared to over 500,000 Mustangs. This disparity highlighted the challenges Plymouth faced in capturing market share in the increasingly crowded pony car segment.

Critical reception of the 1965 Barracuda was generally positive, with automotive journalists praising the car's improved performance and handling, particularly in Formula S guise. Car and Driver magazine wrote, "The Formula S package transforms the Barracuda from a stylish cruiser into a genuine sports car contender. It's a night-and-day difference in terms of handling and performance." However, some critics still felt that the Barracuda's Valiant-based origins held it back in terms of styling distinctiveness compared to purpose-built pony cars.

The 1965 model year represented a significant step forward for the Plymouth Barracuda. The introduction of more powerful engines, the Formula S package, and subtle styling improvements demonstrated Plymouth's commitment to evolving the Barracuda and keeping it competitive in the rapidly changing automotive landscape of the mid-1960s. While it may not have achieved the sales success of its Ford rival, the 1965 Barracuda laid the necessary groundwork for future performance variants. They helped establish the model's reputation among enthusiasts.

Section 2.4: 1966 Model Year Refinements

The 1966 model year marked the final iteration of the first-generation Plymouth Barracuda, and it came with a series of refinements that further enhanced its appeal. As the pony car market continued to heat up, Plymouth made strategic updates to keep the Barracuda competitive and attractive to buyers.

One of the most noticeable changes for 1966 was the updated exterior styling. The Barracuda received a new grille design, giving it a more aggressive and modern look. The front end featured a revised bumper and a more pronounced hood, while the rear saw new taillights that added a touch of sophistication to the car's appearance. These styling tweaks helped to differentiate the 1966 model from its predecessors and keep it fresh in the eyes of consumers.

Under the hood, Plymouth expanded the engine options, offering more power to enthusiasts. The big news for 1966 was the the 273 cu in (4.5L) "Commando" V8 (4-barrel) option. This larger engine significantly boosted the Barracuda's performance capabilities, allowing it to better compete with high-powered versions of the Ford Mustang and Chevrolet Camaro. The 383 V8 produced an impressive 280 horsepower, giving the Barracuda serious muscle car credentials. However, it's worth noting that this engine option came with some compromises, as its size required the removal of power steering and made the car somewhat nose-heavy.

Inside the cabin, Plymouth made several upgrades to enhance comfort and style. New upholstery options were introduced, allowing buyers to customize their Barracudas' interiors to their liking. The dashboard saw minor revisions, improving ergonomics and giving the car a more upscale feel. These interior refinements helped to address some of the criticisms of earlier models and made the Barracuda a more pleasant place to spend time.

Performance enhancements weren't limited to just the new engine option. Plymouth also made improvements to the Barracuda's braking and suspension systems. Upgraded brakes provided better stopping power, which was especially important for the more powerful engine options. The suspension received some tweaks to improve handling and ride quality, striking a better balance between comfort and sportiness.

By the end of its first generation, the Plymouth Barracuda had carved out a unique position in the burgeoning pony car market. While it may not have achieved the same level of sales success as the Ford Mustang, it had developed a loyal following and established itself as a credible performance car. The 1966 model year refinements addressed many of the criticisms of earlier versions and set the stage for the more dramatic changes that would come with the second generation.

The Barracuda's market position at this point was that of an underdog with potential. It offered a unique blend of style, performance, and practicality that appealed to a specific subset of buyers. The fastback design, with its enormous rear window and fold-down seats, provided versatility that some competitors couldn't match. Meanwhile, the introduction of the 383 V8 option signaled Plymouth's intent to be taken seriously in the performance arena.

As the curtain closed on the first generation, the 1966 Plymouth Barracuda stood as a testament to the rapid evolution of the pony car segment. In just three short years, it had grown from a Valiant-based curiosity to a legitimate contender in the performance car market. The refinements made in this final year of the first generation not only improved the car's appeal but also laid the groundwork for the even more impressive second-generation models that were just around the corner.

Section 2.5: Racing and Performance

The first-generation Plymouth Barracuda, born from the humble Valiant, swiftly established itself as a formidable contender in both sanctioned racing events and street performance. From the drag strip to the daily commute, the Barracuda's racing and performance legacy began to take shape during these early years, captivating automotive enthusiasts with its unique fastback design and performance potential.

Early racing efforts for the Barracuda primarily focused on drag racing, where its lightweight design and powerful engine options made it a natural competitor. While Plymouth didn't have an official factory racing program for the Barracuda, several notable private teams and drivers achieved success with modified versions of the car. The introduction of the Formula S package in 1965 provided a solid foundation for racers, with its upgraded suspension and more powerful Commando 273 V8 engine, further enhancing the Barracuda's appeal among performance enthusiasts.

On the street, the Barracuda quickly gained a reputation as a capable performer. Its combination of sleek styling and potent powertrains made it a favorite among young drivers looking for a balance of style and speed. The first-generation Plymouth Barracuda, while often overshadowed by its Ford counterpart, left an indelible mark on American car culture.

Its impact on the automotive landscape was significant, inspiring future generations of muscle cars and pony cars alike. Today, these early Barracudas have become highly sought-after collector's items, their value appreciating as enthusiasts recognize the importance of these pioneering performance machines in the annals of American automotive history.

Section 2.6: The Collectible Cuda

As a collectible, the first-generation Barracuda has seen a steady increase in value over the years. Enthusiasts and collectors appreciate the car's historical significance as one of the earliest entries in the pony car market. The rarity of specific models, particularly those equipped with high-performance options or the Formula S package, has driven prices to impressive heights in recent auctions.

The car's unique features, such as the distinctive wraparound rear window, the largest ever used on a production car at the time, have become major selling points for collectors. This glass masterpiece, which required cutting-edge manufacturing techniques to produce, is often cited as one of the most memorable design elements of any 1960s American car.

Restoration of first-generation Barracudas has become a popular hobby among classic car enthusiasts. The challenge of finding original parts, especially for the more rare configurations, has created a thriving aftermarket industry. Dedicated Barracuda clubs and online communities have emerged, offering resources and support to owners and restorers alike.

The car's influence extends beyond its own model line. The first-generation Barracuda's innovative fastback design and performance-oriented options paved the way for future Plymouth models and influenced competitor designs. Its legacy can be seen in the evolution of the pony car segment and the muscle cars that followed.

Interestingly, the first-generation Barracuda's relative rarity compared to its contemporaries has enhanced its appeal. While it may not have the same widespread recognition as the Mustang, it holds a special place in the hearts of dedicated enthusiasts who appreciate its unique character and historical significance.

Today, a well-preserved or expertly restored first-generation Barracuda is more than just a classic car; it's a rolling piece of automotive history. It represents a pivotal moment in American car design and the birth of a new market segment. For many collectors, owning a first-generation Barracuda is akin to possessing a tangible piece of the American automotive dream, a reminder of an era when style, performance, and innovation converged to create truly iconic vehicles.

Section 2.7: Cultural Impact and Collectibility

The first-generation Plymouth Barracuda, while often overshadowed by its Ford Mustang rival, left an indelible mark on American automotive culture and has since become a highly sought-after collector's item. This section explores the car's cultural significance and its status in the classic car market.

From its introduction, the Barracuda captured the imagination of a generation seeking sporty, stylish transportation. Its unique fastback design and large rear window made it instantly recognizable, setting it apart from other cars on the road. The Barracuda appeared in various forms of popular media, from movies to television shows, cementing its place in the cultural zeitgeist of the 1960s.

One of the most significant cultural impacts of the first-generation Barracuda was its role in popularizing the pony car segment. While the Ford Mustang is often credited with creating this category, the Barracuda's earlier release date gives it the distinction of being the first pony car to market. This fact has become a point of pride among Barracuda enthusiasts and has contributed to the car's legendary status.

The Barracuda also played a part in the evolving youth culture of the 1960s. Its affordable price point and sporty image made it appealing to young buyers, who saw the car as a symbol of freedom and rebellion. The ability to customize and modify the Barracuda also resonated with the DIY spirit of the era, leading to a thriving aftermarket scene that continues to this day.

In terms of collectibility, first-generation Barracudas have seen a steady increase in value over the years. While they may not command the same prices as some of their muscle car contemporaries, they are highly prized by collectors for their historical significance and unique design. The rarest and most desirable models, such as the Formula S package with the 273 Commando V8, can fetch significant sums at auction.

Preservation and restoration of these early Barracudas have become passionate pursuits for many enthusiasts. Owners' clubs and online communities dedicated to the model have sprung up, providing valuable resources for those looking to maintain or restore these classic cars. These groups also organize events and meetups, keeping the spirit of the first-generation Barracuda alive and introducing it to new generations of car enthusiasts.

The car's influence is also evident in modern automotive design. The fastback profile popularized by the Barracuda has made a comeback in recent years, with several manufacturers incorporating similar styling cues into their contemporary models. This resurgence in fastback designs serves as a testament to the enduring appeal of the original Barracuda's silhouette.

In conclusion, the cultural impact of the first-generation Plymouth Barracuda extends far beyond its initial production run. Its unique design, historical significance, and dedicated fan base have ensured its place in the pantheon of classic American automobiles. As these cars become increasingly rare, their value, both monetary and cultural, is likely to continue rising, cementing the first-generation Barracuda's status as a true automotive icon.

Unleashing the Beast: A Comprehensive History of the Barracuda

Chapter 3: Second Generation (1967-1969): Growing Muscles

Section 3.1: The Redesign Revolution

The second generation of the Plymouth Barracuda marked a pivotal moment in the model's history, as Chrysler made the bold decision to break away from the Valiant-based platform that had defined the first generation. This strategic move was driven by the need to compete more effectively in the rapidly evolving pony car market, where rivals like the Ford Mustang and Chevrolet Camaro were gaining ground.

Chrysler's decision to create a unique Barracuda body was a response to the lukewarm reception of the first-generation model. While the original Barracuda had its merits, its close association with the Valiant limited its appeal among performance enthusiasts. Sales figures from the first generation were modest compared to the Mustang, prompting Plymouth to take a more aggressive approach with the redesign.

Unleashing the Beast: A Comprehensive History of the Barracuda

The 1967 Barracuda introduced three distinct body styles, catering to a broader range of preferences. The fastback, which had been the sole option in the first generation, was now joined by a sleek notchback coupe and a stylish convertible.

Each body style had its unique characteristics, with the fastback retaining its sporty profile, the notchback offering a more traditional coupe appearance, and the convertible adding an element of open-air excitement. The fastback remained the most popular choice among buyers, but the additional options helped broaden the Barracuda's appeal. One of the most significant changes in the second generation was the increase in dimensions.

The new Barracuda grew in every direction, with a longer wheelbase, wider track, and more substantial overall proportions. This growth spurt not only improved the car's road presence but also had a profound effect on its performance and handling characteristics. The larger size allowed for better weight distribution and more room for bigger engines, setting the stage for the Barracuda's transformation into a true muscle car.

The enhanced styling cues of the second-generation Barracuda were a dramatic departure from its predecessor. Gone were the awkward proportions and Valiant-derived lines, replaced by a sleek, purposeful design that exuded confidence and aggression. The front end featured a distinctive grille with a delicate mesh pattern, flanked by quad headlights that gave the car a menacing glare. The side profile was characterized by a long hood and short deck, with crisp character lines running the length of the body. Perhaps most notably, the rear-end treatment included unique, full-width taillights that wrapped around the corners, creating an unmistakable nighttime signature.

Inside the cabin, the improvements were equally substantial. The dashboard was completely redesigned, featuring a driver-oriented layout with easy-to-read instruments and controls within easy reach. The seats were more supportive and comfortable, with a range of upholstery options to suit different tastes. Higher-end models offered woodgrain accents and additional amenities, elevating the Barracuda's interior ambiance to new levels of sophistication.

The redesign revolution of the second-generation Barracuda was not just about aesthetics; it was a comprehensive overhaul that touched every aspect of the car. From its more muscular stance to its refined interior, the 1967-1969 Barracuda was a clear statement of Plymouth's intentions to be taken seriously in the pony car wars. This bold reimagining of the Barracuda laid the groundwork for its future success and set the stage for the even more dramatic evolution that was yet to come.

Section 3.2: Performance Enhancements

The second-generation Plymouth Barracuda wasn't just about looking good; it was about going fast. Plymouth engineers understood that to compete in the increasingly competitive pony car market, they needed to inject some serious performance into their redesigned fastback. This section explores the various performance enhancements that transformed the Barracuda from a stylish, compact car into a legitimate muscle car contender.

One of the most significant upgrades came under the hood. Plymouth introduced a range of larger, more powerful engines to satisfy the growing appetite for horsepower. The star of the show was the introduction of the 383 cubic inch (6.3 L) V8 engine. This big-block powerplant was a game-changer for the Barracuda, offering a substantial boost in performance over the previous generation's offerings. With 280 horsepower on tap, the 383-equipped Barracuda could now hold its own against rivals like the Mustang GT and Camaro SS. For those who wanted even more power, Plymouth also offered the 440 cubic inch (7.2 L) V8 in later years, further cementing the Barracuda's muscle car credentials.

But raw power wasn't the only focus. Plymouth also made significant improvements to the Barracuda's suspension and handling characteristics. The car received a revised front suspension geometry, which improved both ride quality and cornering ability. At the rear, a new leaf spring setup helped put the power down more effectively. These changes transformed the Barracuda's driving dynamics, making it more composed and confidence-inspiring on twisty roads. The Formula S package, in particular, offered even more aggressive suspension tuning, catering to enthusiasts who prioritized handling prowess.

Stopping power saw substantial upgrades as well. The second-generation Barracuda featured larger drum brakes as standard, with front disc brakes available as an option. This improvement in braking performance was crucial, given the increased power and weight of the new models. Car and Driver magazine tests from the era showed a notable improvement in stopping distances compared to the first-generation Barracuda, with 60-0 mph braking reduced by nearly 20 feet in some models.

Transmission options were expanded to handle the increased power and cater to different driving preferences. The standard three-speed manual transmission was complemented by an optional four-speed manual transmission for those who desired greater control over gear selection. However, it was the TorqueFlite automatic transmission that proved particularly popular among buyers.

This robust three-speed automatic was well-suited to the Barracuda's newfound muscle, offering smooth shifts and reliable performance. It quickly became the transmission of choice for many Barracuda owners, striking a balance between everyday drivability and the ability to handle high-horsepower applications.

Plymouth also introduced a variety of performance packages and factory options to allow buyers to tailor their Barracudas to their specific needs. The aforementioned Formula S package was the most comprehensive, offering not just the upgraded suspension but also wider tires, special badging, and often the more powerful engine options. Other available performance enhancements included limited-slip differentials, heavy-duty cooling systems, and performance axle ratios. These factory options allowed enthusiasts to create Barracudas that were equally at home on the street or the drag strip.

The culmination of these performance enhancements was evident in the Barracuda's improved acceleration times and overall performance metrics. A well-optioned 1969 Barracuda with the 383 V8 could sprint from 0 to 60 mph in under 7 seconds and complete the quarter-mile in the mid-14-second range, numbers that put it squarely in muscle car territory.

These performance upgrades didn't just improve the Barracuda's capabilities; they fundamentally altered its character. No longer was it merely a compact car with sporty aspirations. The second-generation Barracuda had evolved into an actual performance machine, capable of holding its own against the best Detroit had to offer. This transformation set the stage for the even more potent third generation and cemented the Barracuda's place in muscle car history.

Section 3.3: The Birth of the 'Cuda

The late 1960s marked a pivotal moment in the Plymouth Barracuda's history with the introduction of the 'Cuda name as a performance model. This new designation wasn't merely a marketing gimmick; it represented a significant shift in the Barracuda's identity and capabilities. The 'Cuda emerged as a high-performance variant that would eventually become synonymous with raw power and muscular styling.

The origin of the 'Cuda moniker can be traced back to the everyday nickname enthusiasts had given the Barracuda. Plymouth executives, recognizing the popularity of this shorthand among fans, decided to officially adopt it for their performance-oriented models. This decision proved to be a stroke of marketing genius, as it immediately differentiated the high-performance versions from the standard Barracuda models in the public eye.

The 'Cuda wasn't just a name change; it came with a host of unique features and trim that set it apart from its more pedestrian siblings. Visually, the 'Cuda sported special badging that prominently displayed its new moniker, instantly signaling its performance pedigree. The exterior was further enhanced with unique wheel designs that not only looked more aggressive but also improved the car's handling characteristics. Inside, the 'Cuda received special interior touches, including sport seats with enhanced bolstering, a performance-oriented instrument cluster, and exclusive color and trim options that reinforced its sporting intentions.

Under the hood, the 'Cuda truly shone with its array of high-performance engine options. While the standard Barracuda could be ordered with respectable powerplants, the 'Cuda took things to another level. The star of the show was undoubtedly the 340 cu in (5.6 L) V8, a high-revving small-block that punched well above its weight class.

This engine quickly gained a reputation for its impressive power-to-weight ratio and responsiveness. For those seeking even more brute force, the 383 cu in (6.3 L) V8 was available, offering thunderous acceleration and the kind of low-end torque that muscle car enthusiasts craved.

Plymouth's marketing department wasted no time in positioning the 'Cuda as the crown jewel of their performance lineup. Print ads of the era showcased the 'Cuda's aggressive stance and touted its powerful engine options. Slogans emphasized its performance credentials, with taglines like "The Fish that Swallows Ponies" - a not-so-subtle jab at the Ford Mustang. Television commercials featured the 'Cuda outrunning competitors or performing impressive burnouts, all designed to cement its image as a serious performance machine.

The introduction of the 'Cuda had a profound impact on the Barracuda's image. What was once seen as a somewhat tame, Valiant-based compact was now viewed as a legitimate contender in the muscle car wars. Enthusiast magazines of the time praised the 'Cuda's performance, with many noting that it could hold its own against much more expensive and exotic machinery. Road tests highlighted its impressive acceleration and handling, especially when equipped with the Formula S package.

The 'Cuda variant didn't just improve the Barracuda's performance credentials; it elevated the entire model line in the eyes of the public. Even buyers of standard Barracudas benefited from the halo effect created by the high-performance 'Cuda. The Barracuda was no longer just another pony car; it was now a range that included a serious performance machine capable of running with the best Detroit had to offer.

As the 1960s drew to a close, the 'Cuda had established itself as a force to be reckoned with in the muscle car world. It had transformed the Barracuda from an also-ran in the pony car segment to a genuine performance icon. This transformation set the stage for the even more powerful and legendary 'Cudas that would follow in the next generation, ensuring the Barracuda's place in the pantheon of great American muscle cars.

Section 3.4: Year-by-Year Evolution

The second generation of the Plymouth Barracuda saw significant changes and refinements over its three-year run, each model year bringing its own set of updates and improvements. This evolution not only kept the Barracuda competitive in the rapidly changing pony car market but also helped establish its growing reputation as a serious performance contender.

Unleashing the Beast: A Comprehensive History of the Barracuda

The 1967 model year marked a revolutionary change for the Barracuda. Gone was the Valiant-based design, replaced by a completely new body that was longer, wider, and more muscular in appearance. The redesign was met with enthusiasm from both the automotive press and consumers.

Car and Driver magazine praised the new look, calling it "a quantum leap forward in styling." The increased dimensions allowed for larger engines, including the introduction of the 383 cu in (6.3 L) V8 option, which significantly boosted performance. Sales figures reflected the positive reception, with Barracuda sales jumping to 62,534 units for the model year, a substantial increase over the previous generation's best year.

Moving into 1968, Plymouth made subtle but meaningful updates to the Barracuda. The most noticeable change was a revised grille design, which featured a more pronounced center divider and rectangular parking lights. This gave the front end a slightly more aggressive appearance.

Under the hood, the big news was the introduction of the 340 cu in (5.6 L) V8 as an option. This engine offered an excellent balance of power and weight, making it a favorite among performance enthusiasts. Interior upgrades included new seat designs and additional safety features to comply with federal regulations. Despite increased competition from the likes of the Chevrolet Camaro and Ford Mustang, Barracuda sales remained strong, with 45,412 units sold.

The 1969 model year represented the final refinement of the second generation. Exterior changes were minimal, limited to minor trim adjustments and a redesigned rear valance. However, significant improvements were made under the skin. The suspension was retuned for better handling, and the brake system was upgraded with larger front discs on V8 models.

Unleashing the Beast: A Comprehensive History of the Barracuda

The big news for performance fans was the introduction of the 'Cuda 440 option late in the model year. This limited-production variant was powered by a mighty 440 cu in (7.2 L) V8, making it one of the most powerful pony cars on the market. Despite these improvements, sales dipped slightly to 31,987 units, possibly due to anticipation of the all-new 1970 model.

Throughout the second generation, the Barracuda's sales performance painted an interesting picture. While the initial reception in 1967 was overwhelmingly positive, sales gradually declined over the next two years. This trend was not unique to the Barracuda, as the entire pony car segment faced increased competition and a shifting market. However, it's worth noting that even with the sales decline, the second-generation Barracuda consistently outsold its predecessor, indicating a generally positive market reception.

The critical reception of the second-generation Barracuda was largely favorable. Automotive publications praised its improved styling, increased performance options, and overall refinement compared to the first generation. Motor Trend magazine, in a 1967 review, stated, "Plymouth has transformed the Barracuda from a compact with sporty aspirations into a genuine contender in the pony car wars." The car's handling and balance were frequently cited as strong points, particularly in comparison to some of its muscle-bound competitors.

As the second generation drew to a close in 1969, it had successfully established the Barracuda as a serious player in the pony car market. It had grown from its humble compact car origins into a true performance machine, setting the stage for the even more powerful and iconic third generation that was to follow. The 1967-1969 Barracuda had proven that Plymouth could not only compete in the pony car segment but also innovate and push the boundaries of performance and style.

Section 3.5: Racing and Performance Achievements

The second-generation Plymouth Barracuda didn't just make waves on the streets; it also left its mark on various racing circuits, solidifying its reputation as a performance machine. As the pony car wars heated up, Plymouth recognized the importance of racing success in building brand image and driving sales.

In the world of drag racing, the Barracuda quickly established itself as a formidable competitor. With its potent V8 engines and lightweight design, it became a favorite among both professional racers and weekend warriors. Notable drag racing accomplishments included several class wins at NHRA national events, with the 383-powered Barracudas proving particularly successful in the Super Stock classes. One standout achievement was when a factory-backed Barracuda, piloted by Ronnie Sox, set a new A/Stock Automatic record at the 1968 NHRA Winternationals, showcasing the car's straight-line performance potential.

While not as prominent in stock car racing as some of its competitors, the Barracuda did make appearances in various regional series. Several privateer teams campaigned Barracudas in NASCAR's Grand American series, which was designed for pony cars. Although factory support was limited, these efforts helped demonstrate the Barracuda's durability and competitiveness on oval tracks. In road racing, the Barracuda found its niche in the burgeoning Trans-Am series. The Formula S package, with its 340 cubic inch V8 and upgraded suspension, provided a solid foundation for road racing success. While not as dominant as the Ford Mustang or Chevrolet Camaro in the series, the Barracuda scored several respectable finishes and helped raise the model's profile among road racing enthusiasts.

Land speed records were another arena where the Barracuda sought to make its mark. In 1968, a specially prepared Barracuda, driven by Betty Skelton, set a class land speed record at Bonneville Salt Flats, achieving a two-way average speed of over 180 mph. This accomplishment not only demonstrated the Barracuda's high-speed stability but also highlighted its appeal to female enthusiasts, as Skelton was a renowned female aviator and land speed record holder.

The impact of these racing achievements on the Barracuda's reputation cannot be overstated. Success on the track translated directly to increased interest on the showroom floor. Plymouth's marketing department was quick to capitalize on these accomplishments, featuring racing victories prominently in advertisements and brochures. Slogans like "Win on Sunday, Sell on Monday" became a reality for Plymouth dealers, as performance-minded buyers were drawn to the Barracuda's proven track record.

Moreover, racing success drove continual improvements in the Barracuda's performance capabilities. Lessons learned on the track often found their way into production models, resulting in better handling, more powerful engines, and improved reliability for street cars. This racing-inspired development cycle helped ensure that the Barracuda remained competitive not just on the track but also in the rapidly evolving muscle car market.

The racing achievements of the second-generation Barracuda also played a crucial role in building a loyal enthusiast base. Fans who cheered for the Barracudas on the track often became devoted customers and brand ambassadors, helping to spread the gospel of Plymouth performance through word of mouth and local car clubs.

In retrospect, while the second-generation Barracuda may not have dominated any single racing discipline, its diverse and respectable showing across various motorsports helped establish it as a legitimate performance contender. These racing exploits laid the groundwork for the even more potent third-generation models, setting the stage for the Barracuda's eventual ascension to muscle car royalty. The lessons learned and reputation gained during this period would prove invaluable as Plymouth prepared to unleash the legendary 'Cuda upon the world.

Section 3.6: Special Editions and Rare Models

The second-generation Plymouth Barracuda, although popular in its standard form, also gave rise to several intriguing special editions and rare models that have become highly sought after by collectors and enthusiasts. These limited production variants showcase the versatility and potential of the Barracuda platform, as well as the creativity of Plymouth's marketing team and the ingenuity of aftermarket tuners.

One of the most iconic limited production variants from this era was the Sox & Martin drag race special. Developed in collaboration with the legendary drag racing team of Ronnie Sox and Buddy Martin, these purpose-built Barracudas were stripped down, lightened, and equipped with high-performance engines to dominate the quarter-mile. While only a handful were produced, they played a crucial role in establishing the Barracuda's performance credentials on the drag strip.

Regional special editions also emerged during this period, as Plymouth sought to cater to specific markets and capitalize on local trends. For example, some dealerships in California offered "Beach Barracudas" featuring unique paint schemes, special interior trim, and surf-themed accessories. In the Northeast, dealerships promoted "Winter Warrior" packages that included heavy-duty alternators, upgraded batteries, and even ski racks to appeal to buyers in snowy climates.

While not all concept cars and prototypes make it to production, they often provide fascinating glimpses into what might have been. During the second generation, Plymouth experimented with several intriguing Barracuda concepts. One notable example was the "Barracuda Mako," a sleek, aerodynamic styling exercise that explored potential future design directions. Although it never saw production, elements of its design influenced later Barracuda models and other Chrysler products.

Unleashing the Beast: A Comprehensive History of the Barracuda

For those seeking the ultimate in performance, Plymouth created a limited number of factory experimental models. These purpose-built machines were designed to showcase the full potential of the Barracuda platform and often served as testbeds for new technologies and performance enhancements. One such example was the little-known "B029" program, which produced a handful of lightweight, race-prepped Barracudas featuring exotic materials and hand-built racing engines.

The aftermarket community also embraced the second-generation Barracuda, with numerous companies offering modifications and tuner packages. The most famous of these was Hurst Performance, known for its shifters and performance upgrades. Hurst offered a range of modified Barracudas, including the rare "Hurst Hemi Under Glass" wheelstander exhibition car, which thrilled crowds at drag strips across the country with its crowd-pleasing wheelies.

These special editions and rare models not only added excitement to the Barracuda lineup but also contributed to the car's growing reputation as a serious performance machine. They demonstrated Plymouth's commitment to pushing the boundaries of what was possible with the pony car format. They helped set the stage for the even more outrageous variants that would emerge in the following generation.

Today, these limited-production Barracudas are among the most valuable and sought-after examples of the model. They represent a time when creativity and performance were paramount, and manufacturers were willing to take risks to stand out in an increasingly competitive market. For collectors and enthusiasts, these rare Barracudas offer a tantalizing glimpse into the golden age of American muscle cars, when the only limit seemed to be the engineers' and designers' imagination behind these extraordinary machines.

Section 3.7: Cultural Impact and Influence

The second-generation Plymouth Barracuda left an indelible mark on American culture, extending far beyond its role as a high-performance automobile. As the muscle car era reached its zenith in the late 1960s, the Barracuda found itself at the intersection of pop culture, youth movements, and automotive innovation.

In popular media, the Barracuda made several notable appearances that helped cement its status as an icon of the era. The car's sleek lines and powerful presence made it a natural choice for Hollywood productions looking to capture the essence of the late 1960s. While not as prominently featured as some of its contemporaries, these appearances helped elevate the Barracuda's profile in the public eye.

Celebrity ownership and endorsements played a significant role in shaping the Barracuda's image. Notable figures from the worlds of music, film, and sports were often seen behind the wheel of Plymouth's pony car. For instance, legendary rock musician Jimi Hendrix was known to drive a 1968 Barracuda, adding a touch of counterculture cool to the car's reputation. These high-profile owners helped position the Barracuda as a vehicle of choice for those who wanted to make a statement about their personality and taste.

The impact of the Barracuda on youth culture and the muscle car movement cannot be overstated. As young Americans sought vehicles that reflected their desire for freedom, power, and individuality, the Barracuda offered an appealing alternative to more mainstream options. Its combination of style, performance, and affordability made it accessible to a wide range of buyers, from college students to young professionals. Car clubs dedicated to the Barracuda sprang up across the country, fostering a sense of community among enthusiasts and helping to spread the gospel of Mopar performance.

Unleashing the Beast: A Comprehensive History of the Barracuda

In terms of automotive design trends, the second-generation Barracuda made significant contributions to the field. Its distinctive "coke bottle" styling, characterized by curves that pinched in at the middle of the car before flaring out again at the rear, influenced designers across the industry. The bold use of the fastback profile, particularly in the notchback and convertible variants, helped popularize these body styles among American consumers. Even the Barracuda's unique hood scoops and graphics packages inspired imitators, as other manufacturers sought to capture some of Plymouth's performance mystique.

The legacy of the second-generation Barracuda continues to resonate with automotive enthusiasts today. Its transformation from a compact fastback to a true muscle car contender represents a pivotal moment in the model's history. This generation laid the groundwork for the legendary 1970-1974 models that would follow, establishing the Barracuda as a serious performance machine capable of going toe-to-toe with the best from Ford, General Motors, and other competitors.

Collectors and restorers have come to prize second-generation Barracudas, particularly the high-performance 'Cuda variants, for their historical significance and relative rarity. Values for well-preserved or expertly restored examples have steadily climbed over the years, reflecting the enduring appeal of these vehicles.

Moreover, the second-generation Barracuda's influence can be seen in modern muscle car design. The recent revival of pony cars like the Dodge Challenger (a corporate cousin to the original Barracuda) draws clear inspiration from this era, with styling cues and performance philosophies that pay homage to their 1960s predecessors.

In retrospect, the 1967-1969 Plymouth Barracuda represents more than just a chapter in automotive history. It stands as a testament to an era of American optimism, innovation, and unbridled performance. The cultural impact of this generation of Barracuda helped shape the muscle car narrative and continues to inspire automotive enthusiasm well into the 21st century.

As we look back on this pivotal period in the Barracuda's evolution, we can appreciate how it not only reflected the spirit of its time but also helped define it, leaving an enduring legacy that extends far beyond the realm of cars and into the very fabric of American popular culture.

Chapter 4: Third Generation (1970-1974): The Peak of Pony Car Power

Section 4.1: The All-New 1970 Barracuda

The dawn of the 1970s ushered in a new era for the Plymouth Barracuda, marking a dramatic departure from its humble beginnings as a Valiant-based compact. The 1970 model year saw the introduction of an entirely new Barracuda, one that would forever change the muscle car landscape and cement its place in automotive history.

At the heart of this transformation was the adoption of Chrysler's new E-body platform. This purpose-built chassis was explicitly designed for pony cars, allowing for a wider stance, larger engine bay, and more aggressive styling. The result was a Barracuda that looked and performed like nothing that had come before it.

The new Barracuda's dimensions grew significantly compared to its predecessor. It was now longer, wider, and lower, giving it a more muscular and menacing presence on the road. The increased size wasn't just for show; it allowed for better handling, improved stability, and the ability to house larger, more powerful engines.

Unleashing the Beast: A Comprehensive History of the Barracuda

Gone was the distinctive fastback roofline that had defined the Barracuda since its inception. In its place, Plymouth offered two body styles: a sleek hardtop coupe and a stylish convertible. This move aligned the Barracuda more closely with its pony car competitors, particularly the Mustang and Camaro, while also broadening its appeal to a broader range of buyers.

The redesign wasn't just skin deep. The new E-body platform was shared with Dodge's all-new Challenger, creating a formidable one-two punch in Chrysler's performance car lineup. While the Challenger and Barracuda shared many mechanical components, each maintained its own distinct personality and styling cues, ensuring they wouldn't cannibalize each other's sales.

When the 1970 Barracuda hit showroom floors, the reception was nothing short of electric. Automotive press and muscle car enthusiasts alike were blown away by its bold new look and impressive performance capabilities. Car and Driver magazine declared it "the best handling car ever to come out of Detroit," high praise indeed for a domestic pony car.

Initial sales figures reflected this enthusiasm. In its debut year, Plymouth sold over 55,000 Barracudas, a significant increase over the previous generation's numbers. This success was awe-inspiring given the increasingly crowded pony car market and growing concerns about rising insurance rates for high-performance vehicles.

The all-new 1970 Barracuda represented more than just a new model year update; it was a complete reimagining of what a pony car could be. With its striking design, improved performance, and wide range of options, the third-generation Barracuda had firmly established itself as a major player in the muscle car world. Little did anyone know at the time, but this would be the Barracuda at its peak, a glorious, albeit brief, moment in the spotlight before external forces would begin to dim the muscle car era.

Section 4.2: Powerplant Options: From Mild to Wild

The third-generation Plymouth Barracuda offered an impressive array of engine options, catering to a vast spectrum of drivers, from those seeking economical daily transportation to hardcore performance enthusiasts craving raw power. This diverse lineup of powerplants played a crucial role in establishing the Barracuda's reputation as a versatile pony car capable of meeting various needs and preferences. At the entry level, Plymouth continued to offer the tried-and-true Slant-6 engines.

The base 198 cubic inch (3.2L) Slant-6 provided a modest 125 horsepower, suitable for budget-conscious buyers or those prioritizing fuel economy. For a slight boost in performance, the larger 225 cubic inch (3.7L) Slant-6 was available, producing 145 horsepower. While not particularly exciting, these engines were reliable and economical choices that helped broaden the Barracuda's appeal.

Moving up the performance ladder, the small-block V8 offerings began with the 318 cubic inch (5.2L) engine. This peppy powerplant produced 230 horsepower and offered a noticeable improvement in acceleration and overall performance compared to the Slant-6 options. For those seeking more grunt, the 340 cubic inch (5.6L) V8 was a popular choice, delivering 275 horsepower and providing an excellent balance of power and handling. The 383 cubic inch (6.3L) V8, borrowed from Plymouth's larger cars, rounded out the small-block offerings with 330 horsepower, giving the Barracuda serious muscle car credentials.

However, it was the big-block engines that truly set the third-generation Barracuda apart and cemented its place in the history of muscle cars. The 440 cubic inch (7.2L) V8 was available in two configurations: a four-barrel carburetor version producing 375 horsepower, and the legendary "Six Pack" variant with three two-barrel carburetors, churning out an impressive 390 horsepower. These engines transformed the Barracuda into a formidable straight-line performer, capable of low 14-second quarter-mile times right off the showroom floor.

At the pinnacle of the Barracuda's engine lineup stood the iconic 426 Hemi. This legendary powerplant, with its hemispherical combustion chambers and dual four-barrel carburetors, was conservatively rated at 425 horsepower. In reality, many believed it produced closer to 500 horsepower. Hemi-equipped Barracudas were capable of quarter-mile times in the low 13-second range, making them some of the quickest production cars of their era.

The performance figures of these engines were impressive by any standard. A well-tuned 440 Six Pack Barracuda could sprint from 0 to 60 mph in around 5.8 seconds, while a Hemi-powered version could accomplish the same feat in about 5.6 seconds. These numbers put the Barracuda on par with or ahead of many of its contemporaries, including the Ford Mustang Boss 429 and Chevrolet Camaro Z/28.

When compared to its competitors, the Barracuda's engine lineup stood out for its breadth and top-end power. While Ford's Mustang offered the potent Boss 302 and Boss 429 engines, and Chevrolet's Camaro countered with the Z/28 and COPO options, none could match the raw power of the 426 Hemi in a regular production model. The Barracuda's engine options allowed it to compete at every level of the pony car market, from economical daily drivers to all-out performance machines.

This diverse and potent lineup of engines played a significant role in establishing the third-generation Barracuda, particularly the high-performance 'Cuda variant, as one of the most desirable muscle cars of its era. The availability of these powerplants, especially the big-block options, would later contribute to the car's legendary status and high collectibility among enthusiasts.

Unleashing the Beast: A Comprehensive History of the Barracuda

Section 4.3: The Birth of the 'Cuda

The third generation of the Plymouth Barracuda marked a pivotal moment in the model's history with the introduction of the 'Cuda, a distinct performance variant that would become a legend in its own right. More than just a trim level, the 'Cuda represented a significant leap forward in both style and substance, cementing its place in muscle car lore.

As a stand alone performance model, the 'Cuda was designed to appeal to enthusiasts who craved maximum power and aggressive styling. Standard features included a heavy-duty suspension, performance-tuned exhaust, and a range of high-output engines. The exterior was adorned with bold graphics, including the iconic "hockey stick" stripes that ran along the sides, instantly distinguishing it from its more subdued Barracuda siblings.

The 'Cuda's appearance packages were a feast for the eyes. The standard "performance hood" with dual air intakes was attention-grabbing enough, but options like the shaker hood, which protruded through the hood and visibly shook with the engine's vibrations, took the visual drama to new heights. Bold color options with names like "Lime Light," "In-Violet," and "Tor-Red" ensured that 'Cuda owners would never blend into traffic.

Under the skin, the 'Cuda received significant performance upgrades. The suspension was tightened and reinforced to handle the increased power and improve cornering ability. Larger brakes were fitted to rein in the extra horses, while a limited-slip differential helped put the power to the ground. Engine options ranged from the potent 340 cubic inch small-block to the earth-shaking 426 Hemi, allowing buyers to tailor their 'Cuda to their performance desires and budget.

Plymouth's marketing department recognized the potential of the 'Cuda and positioned it as the brand's halo performance car. Advertisements touted its racing pedigree and street-legal performance, often featuring the car in action on drag strips or winding roads. The tagline "The Beat Goes On" perfectly captured the 'Cuda's blend of muscle and swagger.

The impact of the 'Cuda on Barracuda's reputation and sales was significant. While it represented a smaller percentage of overall Barracuda sales, the 'Cuda drew enthusiasts into showrooms and generated buzz around the entire model line. It elevated the Barracuda from being just another pony car to a genuine performance icon, capable of going head-to-head with the best from Ford, Chevrolet, and Pontiac.

The 'Cuda's influence extended beyond sales figures. It became a staple of car magazine covers, dominating performance comparisons and capturing the imagination of young enthusiasts. Its success on the drag strip, both in stock and modified form, further enhanced its reputation as a serious performance machine.

However, the 'Cuda's time in the spotlight was destined to be brief. As emissions regulations tightened and insurance rates for high-performance cars skyrocketed in the early 1970s, the market for such potent machines began to shrink. Nevertheless, in its short lifespan, the 'Cuda managed to leave an indelible mark on automotive history, becoming one of the most sought-after and valuable muscle cars of all time.

The birth of the 'Cuda represented more than just a new model variant; it was the culmination of Plymouth's performance aspirations, a rolling sculpture that perfectly encapsulated the pinnacle of the muscle car era. Its legacy continues to inspire enthusiasts and collectors, ensuring that the legend of the 'Cuda will live on for generations to come.

Section 4.4: Special Editions and Performance Packages

The third-generation Plymouth Barracuda wasn't just about raw power; it was also about exclusivity and customization. Plymouth offered a range of special editions and performance packages that

Unleashing the Beast: A Comprehensive History of the Barracuda

elevated the Barracuda from a mere muscle car to a collector's dream.

At the forefront of these special editions was the AAR 'Cuda, short for All American Racers. Developed to homologate the Barracuda for Trans Am racing, the AAR 'Cuda was a street-legal version of the race car. It featured a unique 340 cubic inch engine with three two-barrel carburetors, a fiberglass hood with a large air scoop, and side-exit exhaust pipes. The AAR 'Cuda also sported a distinctive paint scheme with a matte black hood and side stripes, making it instantly recognizable. With only 2,724 units produced, the AAR 'Cuda quickly became one of the most sought-after Barracuda variants.

For those seeking luxury alongside performance, Plymouth offered the Gran Coupe option. This package transformed the Barracuda into a more upscale vehicle, featuring premium interior appointments such as high-back bucket seats, wood-grain trim, and plush carpeting. The Gran Coupe appealed to buyers who wanted the power of a muscle car combined with the comfort of a luxury automobile.

Plymouth also tapped into the burgeoning aftermarket scene with its Rapid Transit System (RTS) program. This dealer-installed performance parts catalog allowed Barracuda owners to customize their cars with factory-backed components. The RTS offerings ranged from mild upgrades, like high-performance air cleaners, to more substantial modifications, such as race-ready cylinder heads and camshafts. This program not only boosted performance but also fostered a strong connection between Plymouth and its enthusiast customer base.

One of the most iconic options available on the third-generation Barracuda was the Shaker hood. This functional hood scoop was mounted directly to the engine's air cleaner and protruded through a hole in the hood. As the engine revved, the scoop would visibly shake, hence its name. The Shaker hood became a symbol of muscle car performance and added a distinctive visual element to the Barracuda's already aggressive styling.

Other notable options included the Pistol Grip shifter for manual transmissions, a range of bold colors with names like Lemon Twist and Sassy Grass Green, and the quintessential hockey stick side stripes. These distinctive features allowed buyers to create a Barracuda that was truly their own.

The limited production numbers of many of these special editions and option packages have had a significant impact on their collectibility. For instance, the 1971 Hemi 'Cuda convertible, of which only 11 were produced, has become one of the most valuable muscle cars ever, with examples selling for millions of dollars at auction. Even more common variants with desirable option combinations can command hefty premiums in today's collector car market.

These special editions and performance packages did more than just boost sales; they cemented the Barracuda's place in muscle car history. They demonstrated Plymouth's commitment to performance and personalization, allowing the Barracuda to stand out in an increasingly crowded field of high-powered pony cars. Today, these rare and unique Barracudas serve as rolling time capsules, preserving the pinnacle of American muscle car culture for future generations to admire and enjoy.

Section 4.5: Year-by-Year Changes (1971-1974)

The third-generation Barracuda's peak was short-lived, as changing market conditions and regulations quickly began to impact its design and performance. Each year from 1971 to 1974 brought significant changes, ultimately leading to the model's demise.

In 1971, Plymouth made subtle but noticeable revisions to the Barracuda's appearance. The grille was updated with a more pronounced "frown" design, giving the car a more aggressive look.

The taillights were also modified, now featuring four separate lenses instead of the previous full-width design.

However, the most significant changes occurred under the hood. The process of engine detuning began in response to stricter emissions regulations and rising insurance costs. Compression ratios were lowered across the board, resulting in decreased horsepower ratings for all engine options.

The following year, 1972, saw further power reductions as manufacturers switched to the new SAE net horsepower rating system. This change made the power loss appear even more dramatic on paper, though the actual decrease in performance was less severe than the numbers suggested. More importantly, 1972 marked the last year for two significant Barracuda features: the legendary 426 Hemi engine and the convertible body style.

The loss of these options signaled the beginning of the end for the high-performance era of the Barracuda.In 1973, the impact of new safety regulations became evident in the Barracuda's design. The most notable change was the addition of federally mandated 5 mph front bumpers, which altered the car's sleek profile.

To accommodate these larger bumpers, the grille and front fascia were redesigned, resulting in a heavier and less aesthetically pleasing appearance. The rear of the car received similar treatment in preparation for the 1974 rear bumper requirements.

The final year of production, 1974, saw the Barracuda going out with more of a whimper than a roar. The aforementioned 5 mph rear bumpers were now in place, further compromising the car's styling. Engine options were severely limited, with the most powerful option being a 360 cubic-inch V8 that produced a modest 245 horsepower. The once-mighty 'Cuda was now a shadow of its former self.

Throughout this period, sales steadily declined as various factors contributed to the end of Barracuda's production. The 1973 oil crisis dealt a significant blow to the muscle car market, causing fuel prices to skyrocket and consumer preferences to shift towards more economical vehicles. Stricter emissions standards continued to strangle engine performance, making it increasingly difficult to justify the Barracuda's existence as a performance car.

Insurance companies, wary of the high-powered muscle cars and their young, often inexperienced drivers, raised premiums to prohibitive levels for many potential buyers. This financial barrier further eroded the Barracuda's customer base. Additionally, changing safety regulations required costly redesigns, which were difficult to justify given the model's declining sales. Plymouth found itself in a challenging position, trying to balance performance, compliance, and profitability in a rapidly changing automotive landscape.

By 1974, it became clear that the Barracuda's time had come to an end. The model that had once represented the pinnacle of pony car performance was no longer viable in the new automotive reality. Plymouth made the difficult decision to discontinue the Barracuda, bringing to a close one of the most exciting chapters in American automotive history.

The rapid decline of the third-generation Barracuda from 1971 to 1974 serves as a poignant illustration of how quickly external factors can impact even the most successful automotive designs. It also underscores the fleeting nature of the classic muscle car era, making the early third-generation models all the more special to collectors and enthusiasts today.

Section 4.6: Performance and Handling

The third-generation Barracuda wasn't just about raw power; it was a well-rounded performance machine that could hold its own on both the drag strip and winding roads. Plymouth engineers put considerable effort into refining the car's handling characteristics, resulting in a pony car that was as comfortable carving corners as it was burning rubber in a straight line.

Unleashing the Beast: A Comprehensive History of the Barracuda

At the heart of the Barracuda's improved handling was its suspension setup. The car featured a torsion bar front suspension, which provided a good balance between ride comfort and sportiness. This was paired with leaf springs in the rear, a standard configuration for muscle cars of the era.

However, Plymouth offered a range of suspension upgrades that could transform the Barracuda into a true corner-carver. The performance-oriented 'Cuda models, in particular, benefited from stiffer springs, heavy-duty shock absorbers, and larger sway bars, all of which contributed to reduced body roll and improved cornering stability.

Braking performance was another area where the third-generation Barracuda shone. Base models came equipped with drum brakes all around, which were adequate for everyday driving but could fade under heavy use.

However, Plymouth offered front disc brakes as an option, which significantly improved stopping power and resistance to brake fade. The high-performance 'Cuda models came standard with front discs, and some even offered four-wheel disc brakes for ultimate stopping power. This was a rare feature for American cars of the time and gave the Barracuda a significant advantage over many of its competitors.

Transmission choices played a crucial role in the Barracuda's performance characteristics. The standard three-speed manual transmission was serviceable but not particularly sporty. However, Plymouth offered several upgraded options that significantly enhanced the driving experience. The four-speed manual was a popular choice among enthusiasts, offering crisp shifts and better control over the engine's power band. For those who preferred automatic transmissions, the TorqueFlite three-speed automatic was renowned for its durability and smooth operation. In high-performance applications, it could be remarkably quick-shifting, making it a viable option even for drag racing.

The Barracuda's steering setup was designed to provide good feedback to the driver while maintaining ease of use. Most models used a recirculating ball steering system, which was typical for the era. While not as precise as modern rack-and-pinion setups, it provided a reasonable feel and accuracy. Performance models often featured quicker steering ratios, which enhanced responsiveness and made the car feel more agile in tight turns.

Contemporary road tests and reviews of the third-generation Barracuda were generally positive, particularly for the high-performance variants. Car and Driver magazine praised the 1970 'Cuda 440-6 for its "astonishing acceleration" and noted that its handling was "surprisingly good for a car of its size and weight." Motor Trend was impressed by the car's balanced performance, stating that it "combines drag strip prowess with respectable road manners."

However, reviewers did note some drawbacks. The Barracuda's size and weight, while contributing to its stable highway manners, could make it feel less nimble than some smaller pony cars in tight corners. Additionally, the most powerful engines, particularly the 426 Hemi, could overwhelm the rear tires, making the vehicle challenging to drive at the limit without careful throttle control.

Despite these minor criticisms, the overall consensus was that the third-generation Barracuda, especially in 'Cuda form, was one of the best-performing pony cars of its era. It offered a compelling combination of straight-line speed, cornering ability, and braking performance, making it a favorite among driving enthusiasts. This well-rounded performance package, coupled with its aggressive styling and range of powerful engines, solidified the Barracuda's place in muscle car history and significantly contributed to its enduring appeal among collectors and enthusiasts.

Section 4.7: The Barracuda in Pop Culture

The third-generation Plymouth Barracuda, with its powerful engines and sleek design, didn't just make waves on the streets and

Unleashing the Beast: A Comprehensive History of the Barracuda

drag strips; it also left an indelible mark on popular culture. This iconic muscle car found its way into various forms of media, cementing its status as a symbol of American automotive prowess and youthful rebellion.

In the realm of cinema and television, the Barracuda made numerous memorable appearances. One of the most notable was in the 1971 made-for-TV movie "Duel," directed by a young Steven Spielberg. While the protagonist drove a Plymouth Valiant, a red 1970 Barracuda made a brief but impactful cameo, showcasing the car's aggressive stance and unmistakable profile. The Barracuda also featured prominently in the 1974 film "Gone in 60 Seconds," further solidifying its place in automotive cinema history.

Music, too, embraced the Barracuda's raw power and sexy lines. While Heart's 1977 hit "Barracuda" wasn't directly about the car, its driving rhythm and fierce attitude perfectly captured the spirit of the Plymouth muscle car. Album covers of the era often featured muscle cars, with the Barracuda making several appearances, representing the freedom and rebellion associated with rock and roll.

Celebrities and famous personalities were not immune to the Barracuda's charms. Racing legend Richard Petty, though primarily associated with other Plymouth models, was known to appreciate the Barracuda's performance capabilities. Actor Nicolas Cage, a well-known car enthusiast, has owned several classic muscle cars, including a 1970 Plymouth Hemi 'Cuda, further elevating the car's status among collectors and fans.

In the world of advertising and marketing, Plymouth capitalized on the Barracuda's appeal to youth culture. Magazine ads of the era showcased the car's sporty design and powerful engine options, often featuring young, attractive models to emphasize its connection to the counterculture movement. Dealership brochures highlighted the car's performance credentials, positioning it as the ultimate expression of automotive freedom and individuality.

Unleashing the Beast: A Comprehensive History of the Barracuda

The impact of the third-generation Barracuda on youth culture and car enthusiasm cannot be overstated. It represented a perfect storm of design, performance, and timing, arriving at the peak of the muscle car era when young buyers were eager for vehicles that could express their desire for speed and rebellion. Car clubs dedicated to the Barracuda sprang up across the country, and custom car culture embraced the model, with owners modifying their vehicles for both style and performance.

Even decades after its production ceased, the Barracuda continues to captivate enthusiasts and inspire new generations of car lovers. Its appearances in modern media, including video games like the "Forza" series and "Grand Theft Auto," introduce the classic muscle car to younger audiences, ensuring its legacy lives on. The third-generation Plymouth Barracuda's influence on pop culture serves as a testament to its enduring appeal.

More than just a mode of transportation, it became a rolling sculpture, a performance icon, and a symbol of an era when American muscle ruled the roads. Its presence in movies, music, advertising, and the garages of famous enthusiasts has ensured that the Barracuda will forever remain a celebrated chapter in the annals of automotive history.

Unleashing the Beast: A Comprehensive History of the Barracuda

Chapter 5: Under the Hood: Engines that Defined the Barracuda

Section 5.1: The Early Years: Slant-Six and Small Block V8s

The Plymouth Barracuda's journey to muscle car stardom began with more modest origins. In its early years, the Barracuda was equipped with engines that prioritized reliability and everyday drivability over raw power. This section explores the power plants that laid the foundation for the Barracuda's future performance legacy.

The Slant-Six engine, an inline six-cylinder marvel of engineering, was the base offering for the first-generation Barracuda. Tilted at a 30-degree angle, this unique design allowed for a lower hood line and improved weight distribution. While not a powerhouse by muscle car standards, the Slant-Six earned a reputation for bulletproof reliability and smooth operation. Available in 170 and 225 cubic inch displacements, these engines provided adequate performance for daily driving and helped establish the Barracuda as a practical sports coupe.

In 1964, Plymouth introduced a game-changer for the Barracuda: the 273 cubic inch V8. This small block engine marked the

Barracuda's first step into V8 territory, offering a significant boost in power over the Slant-Six. With its compact dimensions, the 273 fits neatly into the Barracuda's engine bay, maintaining the car's balanced handling characteristics. Initially rated at 180 horsepower, a high-performance version dubbed the "Commando 273" was soon offered, boasting 235 horsepower thanks to a four-barrel carburetor, higher compression, and a more aggressive camshaft.

As the muscle car era gained momentum, Plymouth answered the call for more power with the introduction of the 318 cubic inch V8 in 1967. This engine, an evolution of the 273, provided a notable increase in torque and horsepower while retaining the small block's lightweight characteristics. The 318 quickly became a popular choice among Barracuda buyers, offering a sweet spot of performance and efficiency.

Both the 273 and 318 V8s were receptive to performance modifications, making them favorites among hot rodders and weekend racers. Aftermarket parts manufacturers quickly developed a range of upgrades, from high-flow cylinder heads to performance camshafts, allowing enthusiasts to extract even more power from these versatile engines.

The impact of these early engines on the Barracuda's market position cannot be overstated. While they may not have had the raw power of later big block offerings, these powerplants helped establish the Barracuda as a versatile performer. The combination of the nimble Slant-Six and the increasingly potent small-block V8s allowed the Barracuda to appeal to a wide range of buyers, from economically minded commuters to performance enthusiasts seeking a balanced package.

These early engines laid the groundwork for the Barracuda's future success, proving that the platform could handle increased power and setting the stage for the high-performance variants that would soon follow. As we'll see in the next section, Plymouth was just getting started, and the arrival of the big block engines would propel the Barracuda into the muscle car stratosphere.

Section 5.2: The Birth of the Formula S: 383 Big Block Arrives

The year 1967 marked a significant turning point in the Plymouth Barracuda's history with the introduction of the 383 cubic inch V8 engine. This powerful big block V8 transformed the Barracuda from a sporty compact into a true muscle car contender, giving birth to the legendary Formula S package.

The 383 cu in V8 was a game-changer for the Barracuda. Producing a robust 280 horsepower and 400 lb-ft of torque, it catapulted the car into a new performance category. This engine option allowed the Barracuda to compete head-to-head with other muscle car heavyweights of the era, such as the Chevrolet Camaro SS and the Ford Mustang GT. The 383-equipped Barracuda could sprint from 0 to 60 mph in just under 7 seconds and complete the quarter-mile in approximately 15 seconds, impressive figures for the time.

However, fitting the big block 383 into the Barracuda's engine bay was no small feat. The car's compact A-body platform, initially designed for smaller engines, required significant modifications to accommodate the larger powerplant. Engineers had to redesign the engine mounts, modify the suspension, and rework the exhaust system to accommodate the 383. These changes resulted in some compromises, such as the elimination of power steering in early 383 models due to space constraints.

Despite these challenges, the public and critics alike received the 383-powered Barracuda with enthusiasm. Auto magazines of the day praised its straight-line performance and muscular sound. Car Life magazine declared it "a scrapper of the first order," while Hot Rod called it "a runner with class." The Formula S package, which included the 383 engine along with upgraded suspension and brakes, quickly became the go-to choice for performance enthusiasts.

The introduction of the 383 big block had a profound influence on future Barracuda designs. It paved the way for even larger and more powerful engines in subsequent years, including the legendary 440 and 426 Hemi. The success of the 383 option also pushed Plymouth's designers and engineers to rethink the Barracuda's overall design, leading to the larger E-body platform introduced in 1970, which could more easily accommodate big block engines.

The 383 cu in V8 in the 1967 Barracuda Formula S represented a pivotal moment in the model's evolution. It transformed the Barracuda from a compact pony car into a true muscle car contender, setting the stage for the high-performance variants that would follow and cementing the Barracuda's place in muscle car history. This engine option not only boosted the car's performance credentials but also significantly enhanced its appeal among power-hungry enthusiasts, marking the beginning of the Barracuda's golden age of performance.

Section 5.3: The Golden Age: 440 and 426 Hemi

The late 1960s ushered in the golden age of muscle cars, and Plymouth's Barracuda was not about to be left behind. This era saw the introduction of two legendary engines that would forever cement the Barracuda's place in automotive history: the 440 cubic inch V8 and the iconic 426 Hemi.

The 440 cubic inch V8, first introduced in the Barracuda in 1969, quickly earned a reputation as the "King of the Street." This big-block powerhouse delivered an impressive 375 horsepower and a thunderous 480 lb-ft of torque in its standard form. The 440's combination of massive displacement and relatively simple design made it a favorite among street racers and performance enthusiasts. It offered neck-snapping acceleration and the kind of low-end grunt that could chirp the tires well into second gear.

Unleashing the Beast: A Comprehensive History of the Barracuda

But Plymouth wasn't content to stop there. In 1970, they introduced the 440 Six Pack, a variant featuring three two-barrel carburetors. This setup bumped the power output to 390 horsepower and 490 lb-ft of torque. The Six Pack's triple carbureted madness not only increased performance but also added a visual punch when the hood was popped. The sight of those three carburetors nestled atop the massive 440 was enough to make any gearhead's heart race.

While the 440 was impressive, it was the 426 Hemi that truly became legendary. Nicknamed the "Elephant" due to its massive size and weight, the 426 Hemi was a race-derived engine that found its way into street-legal Barracudas. Officially rated at 425 horsepower and 490 lb-ft of torque, many experts believe these figures were deliberately underrated. In reality, a stock 426 Hemi likely produced between 470 and 500 horsepower.

The Hemi's hemispherical combustion chambers and valve arrangement allowed for excellent airflow and combustion efficiency, resulting in unprecedented power output for a production engine. When properly tuned, a Hemi-equipped Barracuda could rocket through the quarter-mile in the low 13-second range at speeds exceeding 105 mph, numbers that were astounding for the time and remain impressive today.

Performance figures aside, both the 440 and 426 Hemi-equipped Barracudas posted incredible quarter-mile times. A well-driven 440 Six Pack 'Cuda could run the quarter-mile in the mid-13-second range, while the Hemi could dip into the high 12s with an experienced driver. These times made the Barracuda a force to be reckoned with both on the street and at the drag strip.

The rarity and desirability of Hemi-equipped Barracudas cannot be overstated. Due to the engine's high cost and the car's focused performance nature, relatively few were produced. In 1970, the peak year for Hemi 'Cuda production, only 652 hardtops and 14 convertibles were equipped with the 426 Hemi. This scarcity, combined with the engine's legendary status, has made Hemi 'Cudas some of the most valuable and sought-after muscle cars in existence. In recent years, pristine examples have fetched millions of dollars at auction, with the ultra-rare convertibles commanding even higher prices.

The 440 and 426 Hemi engines represented the pinnacle of Mopar muscle. They transformed the Barracuda from a respectable performer into a true street and strip monster. These engines not only defined the Barracuda's performance capabilities but also played a crucial role in shaping its image and legacy. Even today, the mere mention of a Hemi 'Cuda is enough to make any muscle car enthusiast weak in the knees, a testament to the lasting impact these incredible powerplants had on automotive history.

Section 5.4: Small Block Specialists: 340 and 360 cu in V8s

While the big block engines often stole the spotlight, the Barracuda's small block V8s were the unsung heroes that provided an ideal balance of power, weight distribution, and affordability. This section delves into the 340 and 360-cubic-inch V8 engines that became favorites among enthusiasts and racers alike.

The high-revving 340 cu in V8 was introduced in 1968, quickly earning a reputation as a giant-killer. Despite its relatively small displacement, this engine packed a punch that could rival many larger powerplants. With a four-barrel carburetor, the 340 produced an advertised 275 horsepower, though many believe this figure was intentionally underrated. Its lighter weight compared to the big blocks improved handling and allowed for quicker acceleration, making it a popular choice for those who valued agility as much as straight-line speed.

In 1970, Plymouth upped the ante with the introduction of the 340 Six Pack option. This configuration featured three two-barrel carburetors, pushing the official horsepower rating to 290. However, real-world performance suggested that the actual output was closer to 350 horsepower. The Six Pack-equipped Barracudas were formidable opponents on both the street and the strip, often surprising drivers of larger-engined cars.

While the 340 garnered much of the attention, the often-overlooked 360 cu in V8 deserves recognition as well. Introduced in 1971 as a replacement for the 383, the 360 offered a good compromise between the smaller 318 and the larger 440. Although it was initially less potent than the 340 due to lower compression ratios mandated by emissions requirements, the 360 proved to be a robust and tunable engine that would go on to power Mopars for many years to come. When comparing small block performance to their big block counterparts, the results were often surprising.

While the 440 and Hemi engines certainly held the top spots in terms of raw power, the small blocks often provided better real-world performance for the average driver. Their lighter weight allowed for better weight distribution, improved handling, and less strain on other components like brakes and suspension. In many cases, a well-driven 340 Barracuda could outperform its big block siblings on a twisty road or a technical racetrack.

The popularity of these engines among racers and enthusiasts cannot be overstated. The 340, in particular, became a favorite in the SCCA Trans Am racing series, where its combination of power and agility made it highly competitive. In the hands of skilled tuners, these engines could be modified to produce astonishing amounts of power while maintaining reliability.

Enthusiasts appreciated the small blocks for their affordability and ease of maintenance. Parts were plentiful and relatively inexpensive compared to the big blocks, and the engines were easier to work on due to their compact size. This made them ideal for budget-conscious buyers who still wanted impressive performance.

Today, the 340 and 360 engines remain popular choices for Barracuda restorations and resto-mods. Their versatility and potential for high performance, combined with their historical significance, ensure that these small block specialists will continue to be celebrated by Mopar enthusiasts for generations to come. The legacy of these engines serves as a reminder that in the world of muscle cars, bigger isn't always better, and sometimes the most thrilling performances come in smaller packages.

Section 5.5: The Twilight Years: Emissions and the Energy Crisis

As the 1970s progressed, the Plymouth Barracuda, like all muscle cars of the era, faced unprecedented challenges that would ultimately lead to the demise of high-performance V8 engines. The implementation of stringent emissions regulations and the onset of the energy crisis dealt a severe blow to the raw power that had defined the Barracuda's golden age.

The impact of emissions regulations on engine performance was profound and immediate. Automakers, including Plymouth, were required to detune their big-block engines to meet new federal standards. This detuning process involved reducing compression ratios, modifying camshaft profiles, and adjusting ignition timing.

As a result, the once-mighty Barracuda engines began to lose their bite. The 440 cu in V8, once the king of the street, saw its horsepower ratings plummet. Even the legendary 426 Hemi, the crown jewel of Mopar's engine lineup, couldn't escape the regulatory noose, leading to its discontinuation after the 1971 model year.

The introduction of unleaded fuel in the early 1970s further complicated matters for high-performance engines. Designed to reduce harmful emissions, unleaded gasoline was less compatible with the high-compression engines of the muscle car era. This necessitated additional modifications to the Barracuda's powerplants, further reducing their output and efficiency. The switch to unleaded fuel also meant that older Barracudas required modifications to run on the new fuel, creating challenges for owners and enthusiasts.

Unleashing the Beast: A Comprehensive History of the Barracuda

As the energy crisis hit in 1973, it delivered another blow to the Barracuda's powerful engines. With fuel prices skyrocketing and supplies uncertain, consumer demand shifted dramatically towards more fuel-efficient vehicles. The gas-guzzling V8s that had once been the Barracuda's main selling point suddenly became a liability. Plymouth responded by focusing on smaller, more efficient engine options, but this shift came at the cost of the high-performance image the Barracuda had cultivated.

The 1974 model year marked the last hurrah for the Barracuda, and its engine options reflected the changing times. The once-diverse lineup of power plants had been drastically reduced. The top-of-the-line engine was now a shadow of its former self, with performance figures that would have been considered modest just a few years earlier. Even the small block V8s, which had weathered the storm better than their larger counterparts, were shadows of their former selves.

The demise of high-performance V8s signaled the end of an era not just for the Barracuda, but for the entire muscle car genre. The raw, unbridled power that had defined these vehicles was no longer viable in a world increasingly concerned with fuel efficiency and environmental impact. As the last Barracudas rolled off the assembly line, they carried with them the final echoes of a golden age of American automotive performance.

The twilight years of the Barracuda's engine lineup serve as a poignant reminder of how quickly external factors can reshape an industry. In just a few short years, emissions regulations and the energy crisis transformed the automotive landscape, bringing an end to the era of unrestricted horsepower. While this period marked a difficult transition for muscle car enthusiasts, it also set the stage for future innovations in engine technology, as automakers sought to balance performance with efficiency and environmental responsibility.

Section 5.6: Unique Engine Features and Innovations

The Plymouth Barracuda wasn't just about raw power; it also showcased several unique engine features and innovations that set it apart from its competitors. These advancements not only improved performance but also enhanced reliability and efficiency, making the Barracuda a technological leader in its time.

One of the most significant innovations was the introduction of the electronic ignition system. Chrysler pioneered this technology in the early 1970s, and it quickly found its way into the Barracuda's engine bay. This system replaced the traditional points and condenser with a magnetic pickup and electronic control unit. The result was more consistent ignition timing, improved cold starts, and reduced maintenance. For Barracuda owners, this meant fewer tune-ups and more time enjoying their muscle cars on the road.

Another notable feature was the Carter ThermoQuad carburetor, introduced in the later years of the Barracuda's production. This innovative four-barrel carburetor featured a lightweight thermoplastic fuel bowl, which helped reduce heat transfer from the engine. This design incorporated a fuel cooler, preventing vapor lock and enhancing overall performance, particularly in hot weather conditions. The ThermoQuad's large primary bores and even larger secondary bores allowed for improved airflow, contributing to the Barracuda's impressive power output.

High-performance cylinder heads and valvetrains were also crucial components of the Barracuda's most potent engines. The 340 and 440 Six Pack engines, for instance, featured specially designed cylinder heads with larger intake and exhaust valves, as well as improved port designs for enhanced airflow. These heads, combined with high-lift camshafts and heavy-duty valve springs, allowed the engines to rev higher and produce more power than their standard counterparts.

Unleashing the Beast: A Comprehensive History of the Barracuda

The exhaust system designs played a significant role in the Barracuda's performance as well. High-performance models often featured larger diameter exhaust pipes, low-restriction mufflers, and, in some cases, optional factory headers. These systems not only improved exhaust flow but also contributed to the Barracuda's distinctive and aggressive exhaust note. The famous "twin-snorkel" air cleaner, found on many high-performance Barracudas, was another unique feature that enhanced both aesthetics and airflow to the engine.

Factory performance packages offered additional engine modifications that further distinguished the Barracuda. The notorious "Super Stock" package, available on early Hemi-equipped models, included a host of engine tweaks such as heavy-duty connecting rods, special bearings, and modified oiling systems. These modifications were designed to withstand the extreme stresses of drag racing, allowing Barracuda owners to dominate at the strip right off the showroom floor.

One of the most innovative aspects of the Barracuda's engine lineup was the versatility offered to buyers. From economical slant-sixes to fire-breathing Hemis, the range of engine options allowed customers to tailor their Barracuda to their specific needs and desires. This adaptability ensured that whether you were looking for a daily driver or a quarter-mile warrior, there was a Barracuda engine configuration to suit your needs.

These unique engine features and innovations not only contributed to the Barracuda's performance credentials but also showcased Chrysler's engineering prowess. They demonstrated that the Barracuda was more than just a pretty face with a big engine; it was a thoughtfully designed machine that pushed the boundaries of what was possible in a production muscle car. Today, these innovations continue to be celebrated by enthusiasts and serve as a testament to the ingenuity of the era.

Section 5.7: Legacy and Influence

The engines that powered the Plymouth Barracuda throughout its production run left an indelible mark on automotive history, influencing future designs and captivating enthusiasts for decades to come. Their legacy extends far beyond the years when these powerplants roared to life on assembly lines and drag strips across America.

The Barracuda's engines, particularly the high-performance variants, played a crucial role in shaping future Mopar designs. The lessons learned from squeezing massive blocks into the Barracuda's engine bay and the advancements made in small block performance directly influenced the development of subsequent Chrysler, Dodge, and Plymouth vehicles. The 340 small block, for instance, became a favorite among Mopar engineers and found its way into numerous performance models well into the 1970s.

When compared to their Ford and GM contemporaries, the Barracuda's engines often stood out for their raw power and potential. While Ford's 427 and GM's 454 were formidable opponents, nothing quite captured the imagination like Chrysler's 426 Hemi. The Hemi's hemispherical combustion chambers and impressive power output made it a legend both on the street and the track, often giving Barracudas an edge over their muscle car rivals.

The impact of these engines on the hot rodding and restoration community cannot be overstated. Decades after the last Barracuda rolled off the assembly line, enthusiasts continue to seek out and restore these vehicles, with a particular emphasis on preserving or recreating their original powerplants. The 440 Six Pack and 426 Hemi engines remain highly sought after, commanding premium prices and reverence among collectors and restorers alike.

In the world of modern restorations and resto-mods, Barracuda engines remain an enduring inspiration. While some purists insist on numbers-matching originality, many enthusiasts opt for engine swaps that pay homage to the car's performance heritage. It's not uncommon to see restored Barracudas sporting modern Hemi engines, marrying classic styling with contemporary power and reliability. This trend speaks to the enduring appeal of the Barracuda's performance legacy.

The place of Barracuda engines in muscle car folklore is secure. Stories of Hemi-powered 'Cudas outrunning police or dominating at the drag strip have become the stuff of legend, passed down through generations of car enthusiasts. The distinctive rumble of a big block Barracuda has become synonymous with the golden age of American muscle, evoking nostalgia and admiration in equal measure.

Even as the automotive world shifts towards electrification and alternative fuels, the engines that once powered the Barracuda continue to symbolize an era of unrestrained performance and mechanical innovation. They serve as a benchmark against which modern performance cars are often measured, not just in terms of raw power, but in the visceral, emotional response they evoke.

The legacy of Barracuda engines extends beyond mere nostalgia. They represent a high-water mark in American engineering prowess, a time when the quest for performance pushed the boundaries of what was possible with internal combustion. As we look to the future of automotive performance, the spirit of innovation and the pursuit of power that defined the Barracuda's engines continue to inspire engineers and enthusiasts alike.

In the end, the engines that defined the Plymouth Barracuda did more than just power a car; they fueled a passion that continues to burn bright in the hearts of muscle car aficionados around the world. Their influence resonates through time, a testament to an era when the roar of a powerful V8 was music to the ears and the open road was an invitation to unleash American muscle.

Unleashing the Beast: A Comprehensive History of the Barracuda

Chapter 6: Design Evolution: From Valiant-based to Standalone Stunner

Section 6.1: The Valiant Foundation (1964-1966)

The Plymouth Barracuda's journey began with a strong foundation rooted in Chrysler's popular compact car, the Valiant. In 1964, when the Barracuda first hit the showrooms, it was a Valiant with a twist, a sleek fastback design that would set the stage for its future as a muscle car icon.

The decision to base the Barracuda on the Valiant's A-body platform was a strategic move by Chrysler. It allowed for a quick development process and kept production costs low, enabling the company to compete in the rapidly growing pony car market. This shared platform meant that the first-generation Barracuda inherited many of the Valiant's mechanical components and interior elements, including the dashboard, windshield, bumpers, and quarter panels.

However, what truly set the Barracuda apart was its distinctive fastback design. The most striking feature was the massive wraparound rear glass, which at the time was the largest ever installed on a production car. This expansive glass panel, measuring an impressive 14.4 square feet, gave the Barracuda a unique silhouette and greatly improved rear visibility. It also created a spacious, light-filled interior that set it apart from its competitors.

The fastback design was not just about aesthetics; it also served a practical purpose. With the rear seats folded down, the Barracuda offered a cavernous cargo area, making it a versatile option for buyers who wanted both style and functionality. This blend of practicality and sportiness would become a hallmark of the Barracuda's design philosophy.

Despite its Valiant underpinnings, Chrysler's designers worked hard to give the Barracuda its own identity. The front end featured a unique grille design and headlight arrangement, while the rear received special taillights and trim. These subtle but essential differences helped distinguish the Barracuda from its more pedestrian sibling.

The public reception of the first-generation Barracuda was mixed. Some praised its innovative design and versatility, while others criticized it for being too closely related to the Valiant. The car's performance, while respectable, fell short of the standards set by other pony cars of the era, particularly the Ford Mustang, which debuted just two weeks after the Barracuda.

Nevertheless, the first-generation Barracuda laid the necessary groundwork for the model's future. It established the Barracuda as a player in the pony car market and gave Chrysler valuable insights into what buyers in this segment wanted. The lessons learned from this initial offering would prove crucial in shaping Barracuda's evolution in the years to come.

As the 1960s progressed, it became clear that to compete in the muscle car arena truly, the Barracuda would need to differentiate itself from the Valiant further and embrace a more performance-oriented identity. This realization would drive the significant changes seen in the second generation, marking the beginning of the Barracuda's transformation from a Valiant variant to a true muscle car contender.

Section 6.2: Breaking Away from the Valiant (1967-1969)

The second generation of the Plymouth Barracuda marked a significant turning point in its design evolution, as it began to break away from its Valiant roots and forge its own identity. This period, spanning from 1967 to 1969, saw the Barracuda take bold steps towards becoming a true muscle car contender.

The redesign for the 1967 model year introduced a sleeker, more sophisticated roofline that set it apart from its predecessor. Gone was the distinctive wraparound rear glass, replaced by a more conventional but sportier fastback profile. This new silhouette gave the Barracuda a more aggressive stance and improved its aerodynamics, hinting at the performance potential lurking beneath the sheet metal.

The most crucial change in this generation was the expansion of the engine bay. Chrysler's engineers, recognizing the growing demand for more powerful engines in the burgeoning muscle car market, redesigned the Barracuda's front end to accommodate larger powerplants. This foresight would prove invaluable in the coming years, allowing the Barracuda to compete with the likes of the Mustang and Camaro in the horsepower wars.

The front fascia of the second-generation Barracuda underwent a striking transformation. The single headlamps of the previous model were replaced by distinctive quad headlamps, giving the car a wider, more imposing presence on the road. Complementing this new lighting arrangement was a redesigned grille that exuded a sense of performance and aggression. These changes not only enhanced the car's visual appeal but also signaled its growing ambitions in the muscle car segment.

The impact of these design changes on the Barracuda's image cannot be overstated. While the first-generation model was often seen as a hasty response to the Ford Mustang, the second-generation Barracuda began to carve out its own niche. The sportier, more aggressive styling helped shed the image of an economy car associated with its Valiant origins. Car enthusiasts started to view the Barracuda not just as a compact car with performance aspirations, but as a legitimate contender in the muscle car arena.

This evolution in perception was further reinforced by the introduction of the high-performance 'Cuda package in 1969. While not a separate model, the 'Cuda designation signaled the availability of more powerful engines and performance-oriented features, perfectly complementing the car's more athletic design. The second-generation Barracuda's design also allowed for greater customization options, appealing to buyers who wanted to personalize their vehicles. From special paint schemes to performance packages, these options helped broaden the Barracuda's appeal and set the stage for the even more dramatic changes that would come with the third generation.

In essence, the 1967-1969 period represented a crucial transition for the Plymouth Barracuda. Its design evolution during these years reflected Plymouth's growing understanding of the muscle car market and its determination to compete at the highest levels. By breaking away from its Valiant roots and embracing a bolder, more performance-oriented aesthetic, the Barracuda was setting the stage for its transformation into one of the most iconic muscle cars of the early 1970s.

Section 6.3: The 'Cuda Comes into Its Own (1970-1974)

The dawn of the 1970s marked a pivotal moment in the Plymouth Barracuda's history. With the introduction of the all-new E-body platform, the Barracuda finally shed its Valiant-based roots and emerged as a true muscle car contender. This third-generation Barracuda, often referred to simply as the "'Cuda," represented a quantum leap in design and engineering that would cement its place in automotive history.

The E-body platform brought with it a dramatic transformation in the Barracuda's appearance. Gone was the more modest styling of its predecessors, replaced by a wider, lower, and more aggressive stance that commanded attention on the street. This new Barracuda was a full five inches wider than its predecessor, giving it a more planted and muscular appearance. The lower roofline and longer hood created a sleeker profile that exuded speed even when the vehicle was standing still.

The most iconic feature of the third-generation Barracuda was its distinctive front-end design. The "cheese grater" grille, characterized by its series of narrow, horizontal openings, became an instant classic. Flanked by quad headlamps, this aggressive front fascia gave the Barracuda a menacing presence on the road. The design was not just for show; it also improved cooling for the larger engines that now fit comfortably under the expansive hood.

The new E-body platform allowed Plymouth to offer the Barracuda in three distinct body styles, each with its own appeal. The hardtop coupe provided a classic muscle car silhouette, while the convertible offered open-air thrills. However, it was the fastback that truly captured the essence of the Barracuda's sporting heritage, with its sleek, aerodynamic profile harking back to the original 1964 design while bringing it firmly into the 1970s.

This design overhaul did more than just turn heads; it fundamentally altered the Barracuda's image and market position. No longer was it seen as a compact car with performance aspirations. The third-generation Barracuda was now a full-fledged muscle car, ready to go toe-to-toe with the likes of the Ford Mustang, Chevrolet Camaro, and even its corporate cousin, the Dodge Challenger.

The new design also allowed for a broader range of engine options, including the legendary 426 Hemi and 440 Six Pack, further solidifying the Barracuda's performance credentials. The wider body and larger engine bay meant that these powerful motors could be easily accommodated, turning the 'Cuda into a high-performance machine.

The impact of this design evolution on the Barracuda's status in the muscle car pantheon cannot be overstated. It transformed from a respected but somewhat overlooked player into a bona fide icon of the muscle car era. The aggressive styling, combined with potent performance options, made the third-generation Barracuda a dream car for enthusiasts and a nightmare for competitors on both the street and the track.

Moreover, the distinctive design of the 1970-1974 Barracuda has stood the test of time, remaining highly sought after by collectors and enthusiasts today. Its bold lines and aggressive stance continue to evoke the golden age of American muscle cars, serving as a testament to the power of innovative design in creating automotive legends.

In essence, the third-generation Plymouth Barracuda represented the culmination of its design evolution. It was the moment when the 'Cuda truly came into its own, shedding its economy car origins and emerging as a standalone stunner that could hold its own against any muscle car of its era. This transformation from Valiant variant to muscle car icon was complete, and the automotive world would never be the same.

Section 6.4: Exterior Design Elements

The evolution of the Plymouth Barracuda's exterior design is a fascinating journey that reflects the changing aesthetics and performance demands of the muscle car era. One of the most striking transformations occurred in the front fascia across generations. The first-generation Barracuda, with its Valiant-derived styling, featured a simple grille and headlight arrangement. As the model progressed, the front end became more aggressive and distinctive, culminating in the third generation's iconic "cheese grater" grille and menacing quad headlamps that gave the car a predatory appearance.

Hood designs also underwent significant changes throughout the Barracuda's lifespan. Early models featured relatively flat hoods, but as performance became a greater focus, more aggressive hood designs were introduced. Performance hood options, such as the shaker hood and the power bulge hood, not only enhanced the car's visual appeal but also served functional purposes, accommodating larger engines and improving air intake. These muscular hood designs became signature elements that set the Barracuda apart from its competitors.

The Barracuda's side profile and character lines evolved dramatically over time. The first-generation model's fastback design was revolutionary for its time, featuring a distinctive wraparound rear window. As the model progressed, the side profile became sleeker and more sculpted. The second generation introduced a more pronounced hip line, while the third generation E-body platform allowed for a wider, lower stance with more pronounced wheel arches and a longer hood, giving the car a more powerful and athletic appearance.

Rear styling saw the most dramatic changes of all. The first generation's fastback design with its unique wraparound glass was a defining feature. However, as the model evolved, Plymouth moved towards more traditional designs. The second generation maintained a fastback profile but with a more conventional rear window. The third generation offered multiple body styles, including a more traditional coupe design, a convertible, and a revised fastback. Each iteration refined the rear styling, with the final generation featuring a distinctive rear valance and taillights that emphasized the car's width and low-slung stance.

Throughout its evolution, aerodynamics played an increasingly important role in the Barracuda's design decisions. While early models were styled primarily for aesthetic appeal, later generations incorporated more aerodynamic considerations. The third-generation Barracuda, in particular, featured a more wind-cheating profile with its sloping nose and fastback design. These aerodynamic improvements not only enhanced the car's performance capabilities but also contributed to its sleek, modern appearance.

The exterior design elements of the Plymouth Barracuda tell a story of a car that grew from a modest compact into a full-fledged muscle car icon. Each generation built upon the last, refining and reimagining what a performance car could look like. From its front fascia to its rear styling, every aspect of the Barracuda's exterior was carefully crafted to create a vehicle that was both visually striking and performance-oriented. This evolution in design not only reflected the changing tastes of the era but also played a crucial role in establishing the Barracuda as one of the most recognizable and beloved muscle cars of all time.

Section 6.5: Interior Design Evolution

The evolution of the Plymouth Barracuda's interior design is a fascinating journey that mirrors the car's overall transformation from a modest compact to a full-fledged muscle car icon. This progression not only reflected changing automotive trends but also the shifting expectations of performance car enthusiasts.

In its first generation, the Barracuda's interior was a clear reflection of its Valiant roots. The cabin was modest and utilitarian, with a straightforward dashboard layout that prioritized function over form. The seating arrangement was typically bench-style, common in economy cars of the era. While comfortable, it lacked the sporty feel that would later become synonymous with muscle cars. The instrument cluster was simple, featuring basic gauges that provided essential information without frills. This interior, while practical, did little to set the Barracuda apart from its more pedestrian counterparts.

As the Barracuda entered its second generation, Plymouth began to inject more sportiness into the interior design. Recognizing the car's potential as a performance vehicle, designers introduced elements that appealed to enthusiasts. Bucket seats became an option, offering better lateral support during spirited driving. The dashboard received a more driver-centric layout, with gauges angled slightly towards the driver for better visibility. Chrome accents and woodgrain trim options added a touch of sophistication, elevating the interior ambiance. These changes, while subtle, signaled the Barracuda's gradual shift towards a more performance-oriented identity.

The third generation of the Barracuda saw the most dramatic transformation in interior design. With the introduction of the E-body platform, Plymouth fully embraced the muscle car ethos in the cabin. The most notable change was the adoption of a driver-focused cockpit design. This layout wrapped around the driver, creating a more immersive and connected driving experience. The instrument panel was redesigned with a sportier look, featuring deep-set gauges that were easy to read at a glance. Tachometers and auxiliary gauges became more prominent, catering to performance-minded drivers who wanted to monitor their engine's vital signs.

Seating in the third-generation Barracuda underwent significant evolution. Bucket seats became standard, offering improved support and comfort during high-speed maneuvers. These seats were often available with high-back designs and a variety of upholstery options, including performance-inspired materials like vinyl and leather. The rear seats in hardtop and fastback models were designed to fold down, increasing cargo space and versatility.

The evolution of the dashboard and instrument cluster designs throughout the Barracuda's lifespan is particularly noteworthy. From the simple, flat dash of the first generation, the design progressed to a more contoured and ergonomic layout. In the final iteration, the dashboard featured a distinctive dual-cowl design, with a separate section for the passenger side. This not only looked sportier but also improved safety by providing a padded area in front of the passenger.

The instrument cluster underwent a similar transformation. Early models featured basic analog gauges for speed, fuel, and engine temperature. Later models, especially in high-performance trims, boasted comprehensive instrument panels with additional gauges for oil pressure, battery charge, and even engine vacuum. The introduction of rally-style instrumentation in some models further emphasized the Barracuda's performance credentials.

Throughout its evolution, the Barracuda's interior also saw improvements in materials and build quality. While early models relied heavily on vinyl and basic fabrics, later generations incorporated higher-quality materials, including better grades of vinyl, leather, and more durable carpeting. Attention to detail also improved, with a better fit and finish and more thoughtful ergonomics. The interior design evolution of the Plymouth Barracuda is a clear reflection of its journey from a Valiant variant to a muscle car icon. Each generation brought significant improvements, not just in style but in functionality and driver engagement.

This progression demonstrates Plymouth's commitment to creating a car that was not just powerful on the outside, but also delivered a thrilling and comfortable experience on the inside. The final iteration of the Barracuda's interior stood as a testament to the golden age of muscle cars, where performance, style, and driver-focused design converged to create truly special automobiles.

Section 6.6: Color and Trim: Making a Statement

The Plymouth Barracuda's evolution wasn't just about shape and form; color and trim played a crucial role in defining its identity and appeal. Throughout its production run, the Barracuda's color palette and trim options evolved to reflect changing tastes, cultural shifts, and the car's transition from modest compact to muscle car icon.

In the realm of muscle cars, color was more than just a visual preference; it was a statement. The Barracuda's color options evolved dramatically across its three generations, mirroring the car's own transformation. In its early years, when it was still closely tied to the Valiant, the Barracuda's color choices were relatively conservative, featuring subdued tones that appealed to a broad audience. These included classic options like white, black, and various shades of blue and green.

As the Barracuda began to establish its own identity in the second generation, its color options expanded to include more vibrant hues. This shift reflected the growing youth market and the increasing association of the Barracuda with performance and style. Brighter blues, reds, and yellows began to appear, giving buyers more opportunity to express themselves through their car's appearance.

However, it was in the third generation that the Barracuda's color options truly exploded with personality. This era introduced what Chrysler called "High Impact Paint" colors, a range of bold, eye-catching hues that perfectly complemented the car's muscular new design. These colors weren't just paint options; they were part of the Barracuda's identity and appeal. With names like "Plum Crazy," "Sublime," "Go Mango," and "In-Violet," these colors were as much a part of the muscle car experience as the roar of a V8 engine.

The impact of these vivid color choices cannot be overstated. They allowed each Barracuda to stand out in a crowd, reflecting the individuality of its owner. These bold colors also served to emphasize the car's curves and aggressive stance, making the Barracuda even more visually striking.

Complementing these attention-grabbing colors were a variety of stripe and decal packages that further enhanced the Barracuda's appearance. These ranged from simple side stripes to more elaborate graphics, often tied to specific performance packages. The iconic "hockey stick" stripe, which ran along the side of the car and hooked upward at the rear, became a signature element of many Barracudas. Other popular options included hood stripes, rear panel blackout treatments, and model-specific decals like the "AAR" graphics on the All American Racers edition.

These color and trim choices weren't just about aesthetics; they reflected the era and target market of the Barracuda. The transition from subdued colors to bold, almost psychedelic hues mirrored the cultural shifts of the 1960s and early 1970s. The youth market, which was a key demographic for muscle cars, gravitated towards these expressive color options. It allowed them to make a statement, to stand out, and to showcase their personality through their vehicle choice.

Moreover, these vibrant colors and distinctive trim packages served a marketing purpose. They made the Barracuda instantly recognizable on the street and in advertisements, helping to build the car's brand identity. The High Impact Paint colors, in particular, have become so closely associated with Chrysler's muscle cars that they've become collectible in their own right, with specific color combinations commanding premium prices in the classic car market today.

In essence, the evolution of the Barracuda's color and trim options tells a story parallel to that of the car itself. From modest beginnings to bold statements, the paint and graphics offered on the Barracuda reflected its journey from economy car variant to muscle car legend. These choices allowed the Barracuda not only to perform like a muscle car but also to look the part, cementing its place in automotive history as a true style icon.

Section 6.7: Design Influences and Inspirations

The Plymouth Barracuda's design evolution was not created in a vacuum. It was shaped by a complex interplay of influences, inspirations, and external factors that collectively guided its transformation from a modest fastback to a muscle car icon.

Contemporary muscle cars played a significant role in influencing the Barracuda's design trajectory. As competitors like the Ford Mustang and Chevrolet Camaro gained popularity, Plymouth designers took note of the elements that resonated with consumers.

Unleashing the Beast: A Comprehensive History of the Barracuda

The long hood, short deck proportions that became synonymous with pony cars were embraced and refined in the Barracuda's later generations. The aggressive stance and bold styling cues of vehicles like the Pontiac GTO also left their mark, inspiring Plymouth to push the Barracuda's design in a more muscular direction.

Racing requirements were another crucial factor in shaping the Barracuda's design. As Plymouth sought to establish the car's performance credentials, certain design elements were influenced by the needs of the track. The wider, lower stance of the third-generation Barracuda was partly a response to the demands of high-speed stability. The expansive hood scoops and aerodynamic enhancements were not just for show; they served practical purposes in racing applications. The need for engine bay space to accommodate larger, more powerful engines also dictated confident design choices, particularly in the transition to the E-body platform.

Changing safety regulations had a profound impact on automotive design during the Barracuda's lifetime, and the car's evolution reflects this. The transition from the distinctive wraparound rear glass of the first generation to more conventional designs in later models was partly influenced by safety considerations. The introduction of required safety features, such as side marker lights and impact-absorbing bumpers, in the late 1960s and early 1970s necessitated design adaptations, challenging designers to incorporate these elements while maintaining the car's aesthetic appeal.

Key designers played a crucial role in shaping the Barracuda's distinctive look. While many hands contributed to the car's design over the years, specific individuals left an indelible mark. John E. Herlitz, who later became Chrysler's vice president of design, was instrumental in crafting the third-generation Barracuda's aggressive, muscular appearance. His vision helped transform the Barracuda from a competent pony car into a true muscle car icon.

Consumer preferences were always at the forefront of design decisions for the Barracuda. As tastes evolved from the more conservative styles of the early 1960s to the bolder, more expressive designs of the late 1960s and early 1970s, the Barracuda evolved to meet these changing preferences. The introduction of the "High Impact" color palette, for instance, was a direct response to consumers' desire for more eye-catching, individualistic vehicles.

Broader cultural trends also influenced the Barracuda's design. The optimism and exuberance of the 1960s, coupled with the fascination with speed and power, found expression in the car's increasingly bold and aggressive styling. As the muscle car era reached its zenith, the Barracuda's design reflected the zeitgeist, embodying the spirit of youthful rebellion and American automotive prowess.

In essence, the Plymouth Barracuda's design evolution was a dynamic process, responsive to a myriad of influences. From the competitive pressures of the muscle car market to the practical demands of racing, from changing safety standards to the visionary ideas of key designers, and from evolving consumer tastes to broader cultural trends, all these factors combined to shape the Barracuda's iconic design. The result was a car that not only reflected its era but also pushed boundaries, ultimately securing its place in the pantheon of classic American muscle cars.

Unleashing the Beast: A Comprehensive History of the Barracuda

Chapter 7: Racing Pedigree: The Barracuda on Track and Strip

Section 7.1: The Barracuda's Early Racing Days

The Plymouth Barracuda's journey into the world of competitive racing began shortly after its introduction in 1964. As the automotive landscape of the 1960s was dominated by performance and speed, Plymouth recognized the importance of establishing a strong racing presence for its new pony car. The Barracuda's early racing days were marked by a mix of ambitious goals, steep learning curves, and the determination to carve out a place in the highly competitive world of motorsports.

Plymouth's racing ambitions for the Barracuda were clear from the start. The company saw racing not just as a way to prove the car's performance capabilities, but also as a crucial marketing tool to appeal to the speed-hungry youth market of the era. This strategy, often summarized as "Win on Sunday, Sell on Monday," was a driving force behind Plymouth's push to get the Barracuda onto racetracks as quickly as possible.

Unleashing the Beast: A Comprehensive History of the Barracuda

The Barracuda's first forays into competitive racing between 1964 and 1966 were met with mixed results. Initially, the car struggled to make a significant impact, particularly when pitted against more established competitors like the Ford Mustang and Chevrolet Camaro. The Barracuda's unique fastback design, while visually striking, presented some aerodynamic challenges on high-speed circuits. Additionally, the early engine options, including the 273 cubic inch V8, were often outmatched by the more powerful offerings from rival manufacturers.

Despite these initial setbacks, some notable early racers saw potential in the Barracuda and began modifying the cars to improve their performance. One such pioneer was Don Garlits, the legendary drag racer, who campaigned a modified 1965 Barracuda in various drag racing events. Garlits' Barracuda, equipped with a highly tuned Hemi engine, helped demonstrate the platform's potential and inspired other racers to give the Barracuda a chance.

Another significant figure in the Barracuda's early racing history was Akron Arlen Vanke, who achieved success with his "Akron Arlen" Barracuda in NHRA drag racing events. Vanke's modifications, which included significant weight reduction and engine tuning, helped showcase the Barracuda's capabilities when pushed to its limits.

However, these early racing efforts were not without their challenges. The Barracuda faced stiff competition from more established models, and its relative newness meant that there was a lack of readily available performance parts and racing knowledge specific to the platform. Many early racers had to rely on trial and error, often developing custom solutions to enhance the Barracuda's performance.

The handling characteristics of the early Barracudas also presented challenges on road racing circuits. The car's weight distribution and suspension geometry required significant modification to make it competitive against purpose-built racing machines. This led to extensive experimentation with suspension setups, weight reduction techniques, and aerodynamic changes.

Despite these obstacles, the lessons learned during this period proved invaluable for the future development of the Barracuda. The feedback from racers and the data gathered from competition directly influenced Plymouth's engineering decisions for subsequent model years. Engineers paid close attention to the modifications being made by successful racers, incorporating many of these ideas into factory performance packages and later production models.

The early racing experiences also highlighted the importance of a robust engine program. This led to the development and introduction of more powerful engine options in later years, including the 340 and 440 cubic inch V8s, as well as the legendary 426 Hemi. Furthermore, the challenges faced in these early years fostered a tight-knit community of Barracuda racers and enthusiasts. This grassroots support would prove crucial in the years to come, as knowledge and experience were shared, helping to accelerate the Barracuda's development as a serious contender on the track.

In retrospect, the Barracuda's early racing days, while not always victorious, laid a crucial foundation for future success. The lessons learned, the modifications developed, and the passion ignited during this period set the stage for the Barracuda to evolve into a formidable force in various racing disciplines. These early efforts paved the way for the impressive achievements that would follow, cementing the Barracuda's place in racing history and contributing significantly to its enduring legacy in the world of muscle cars.

Section 7.2: Drag Racing Dominance

The Plymouth Barracuda's ascent to drag racing stardom was nothing short of meteoric. As the muscle car era hit its stride in the mid-1960s, the quarter-mile became the ultimate proving ground for Detroit's finest, and the Barracuda was more than ready to answer the call. The Barracuda's popularity in drag racing circles skyrocketed, thanks in large part to its potent engine options and lightweight design. Racers quickly recognized the potential of the compact A-body platform, which provided an excellent foundation for building a competitive drag car.

The introduction of the 383 cubic inch V8 in 1967 and the legendary 426 Hemi in 1968 catapulted the Barracuda into the upper echelons of drag racing performance. To achieve success on the drag strip, Barracuda racers implemented a series of key modifications. Weight reduction was paramount, with teams stripping interiors, replacing body panels with fiberglass components, and even acid-dipping entire bodies to shed precious pounds.

Suspension upgrades, including lightweight torsion bars and special rear leaf springs, helped improve weight transfer and traction off the line. Engine modifications ranged from high-flow cylinder heads and aggressive camshafts to exotic multi-carburetor setups and superchargers for the most extreme builds.

The Barracuda's drag racing legacy was built on the backs of legendary racers who piloted these machines to victory. Names like Sox & Martin, Dick Landy, and Ronnie Sox became synonymous with Barracuda drag racing success. Sox & Martin, in particular, dominated the Super Stock classes with their Hemi-powered Barracudas, setting numerous records and clinching multiple championships throughout the late 1960s and early 1970s.

Record-breaking runs and notable victories soon followed. In 1968, Ronnie Sox piloted his Barracuda to win in the NHRA Springnationals, setting a new A/Stock Automatic record with a blistering 10.87-second quarter-mile run. The following year, at the 1969 NHRA Winternationals, Sox shattered his own record with a 10.75-second pass, firmly establishing the Barracuda as a force to be reckoned with in drag racing.

The impact of these drag racing successes on the Barracuda's street reputation cannot be overstated. The halo effect of drag racing glory helped cement the Barracuda's image as a true muscle car contender, capable of going toe-to-toe with the best from Ford and General Motors. Moreover, the lessons learned on the drag strip directly influenced the development of street-legal Barracudas. Plymouth engineers incorporated many of the racing-derived improvements into production models, resulting in cars that were more powerful, better handling, and more capable of high-performance driving right off the showroom floor.

As the 1960s gave way to the 1970s, the Barracuda continued to evolve, with each new model year bringing advancements that kept it competitive both on the street and the strip. The introduction of the E-body Barracuda in 1970, with its wider stance and ability to accommodate larger engines, only served to enhance its drag racing prowess.

The Barracuda's drag racing dominance not only solidified its place in muscle car history but also created a legacy that continues to inspire enthusiasts and racers to this day. From smoky burnouts to lightning-fast quarter-mile runs, the Barracuda's time on the drag strip remains one of the most exciting chapters in its storied history.

Section 7.3: Trans-Am Series and Road Racing

The Barracuda's racing prowess wasn't limited to the straight-line battles of the drag strip. Plymouth's pony car also made its mark in the challenging world of road racing, most notably in the highly competitive Trans-Am Series. This arena provided a perfect stage for the Barracuda to showcase its versatility and prove that it could hold its own against the best in high-speed cornering and endurance.

Plymouth's entry into the Trans-Am Series marked a significant milestone in the Barracuda's racing history. The decision to compete in this series was driven by a desire to demonstrate the car's all-around performance capabilities and to challenge rivals like the Ford Mustang and Chevrolet Camaro on twisting road courses. This move required extensive modifications to the Barracuda, transforming it from a street-oriented muscle car into a nimble road racing machine.

To prepare the Barracuda for the unique demands of road racing circuits, Plymouth's engineers had to rethink many aspects of the car's design. Suspension systems were completely overhauled, with stiffer springs, heavy-duty shock absorbers, and reinforced anti-roll bars to improve cornering stability. The braking system saw significant upgrades to handle the repeated high-speed deceleration required in road racing. Aerodynamic improvements, including front air dams and rear spoilers, were added to increase downforce and stability at high speeds.

The heart of these road racing Barracudas was their engines. While displacement was limited by series regulations, Plymouth's engineers squeezed every ounce of power they could from their small-block V8s. Advanced fuel injection systems, high-flow cylinder heads, and carefully tuned exhaust systems helped these engines rev higher and produce more power than their street counterparts.

Several notable drivers and teams campaigned Barracudas in the Trans-Am Series. Among the most prominent was Dan Gurney's All-American Racers team. Gurney, a legendary figure in American motorsports, brought his wealth of experience to bear in developing and racing the Barracuda. Other skilled drivers, such as Swede Savage and Sam Posey, also took turns behind the wheel, each contributing to the car's development and on-track success.

Both triumphs and challenges marked the Barracuda's journey in the Trans-Am Series. While it faced stiff competition from more established players, the Barracuda managed to score several impressive victories and podium finishes. These successes were hard-fought, often coming down to skilled driving, clever strategy, and the Barracuda's ability to maintain pace over long races.

One of the most significant challenges the Barracuda faced in Trans-Am racing was reliability. The intense nature of road racing put tremendous stress on every component of the car. Plymouth's engineers worked tirelessly to address weak points, constantly evolving the design to withstand the rigors of competition. This process of continuous improvement not only benefited the race cars but also influenced the development of production Barracudas.

The Barracuda's participation in road racing had a profound impact on its handling and performance characteristics. Lessons learned on the track directly influenced improvements in the streetcars. The suspension tuning, brake upgrades, and aerodynamic enhancements developed for racing found their way, in modified form, into production Barracudas. This racing pedigree gave the street cars a level of handling prowess that set them apart from many of their muscle car contemporaries.

Moreover, the Barracuda's presence in the Trans-Am Series significantly enhanced its performance image. No longer was it seen solely as a straight-line speedster; the Barracuda had proven its ability to carve corners with the best of them. This well-rounded performance profile appealed to a broader range of enthusiasts, helping to cement the Barracuda's reputation.

The legacy of the Trans-Am Barracudas extends far beyond their on-track achievements. These cars represented the pinnacle of Plymouth's engineering capabilities, showcasing what was possible when the company's best minds focused on creating a world-class racing machine. Today, surviving Trans-Am Barracudas are highly prized by collectors, serving as tangible links to an era when American pony cars battled for supremacy on road courses across the nation. In the grand narrative of the Plymouth Barracuda, its foray into road racing stands as a testament to the car's versatility and the ambition of its creators. From the high-banked turns of Daytona to the challenging curves of Laguna Seca, the Barracuda proved it was more than just a muscle car; it was an actual performance machine capable of excellence in any racing discipline.

Section 7.4: NHRA Super Stock and Pro Stock Classes

The National Hot Rod Association (NHRA) provided a thrilling stage for the Plymouth Barracuda to showcase its raw power and speed. As the muscle car era reached its zenith, the Barracuda found itself in the thick of the action in both the Super Stock and Pro Stock classes, leaving an indelible mark on drag racing history.

The Barracuda's presence in NHRA competitions was nothing short of spectacular. From the moment it entered these highly competitive classes, it became clear that Plymouth had created a machine capable of going toe-to-toe with the best in the business. The thunderous roar of its high-performance engines and the sight of its sleek form hurtling down the quarter-mile became a familiar and welcome sight for drag racing enthusiasts.

The development of Super Stock and Pro Stock Barracudas was a testament to Plymouth's commitment to racing excellence. For the Super Stock class, engineers pushed the boundaries of what was possible with a production-based vehicle. They fine-tuned every aspect of the car, from its suspension geometry to its weight distribution, creating a purpose-built drag racing machine that still maintained a connection to its street-going counterparts.

The Pro Stock class, introduced in 1970, presented a new challenge and opportunity for the Barracuda. This class allowed for even more extensive modifications, and Plymouth rose to the occasion. The Pro Stock Barracudas were technological marvels, incorporating cutting-edge innovations that would influence drag racing for years to come.

Key technological innovations for NHRA racing included the development of the iconic Hemi engine, which became synonymous with Chrysler's drag racing efforts. The Hemi's hemispherical combustion chambers and efficient valve arrangement enabled superior breathing and combustion, resulting in incredible horsepower figures. Additionally, advanced lightweight materials, aerodynamic modifications, and specialized transmission systems were employed to give the Barracuda every possible advantage on the drag strip.

The Barracuda's NHRA efforts were piloted by some of the most legendary names in drag racing history. Drivers like Sox & Martin, Dick Landy, and Don Grotheer became household names among racing fans, their skills behind the wheel matched only by the capabilities of their Barracudas. These drivers pushed their machines to the limit, setting records and clinching victories that would be talked about for generations.

Perhaps the most famous of all was "The Goldfish," a Barracuda campaigned by Sox & Martin. This iconic car, with its gleaming gold paint job and unmatched performance, became a symbol of Plymouth's dominance in the Pro Stock class. Its numerous victories and record-breaking runs cemented the Barracuda's status as a drag racing legend.

The impact of NHRA success on the Barracuda's performance image cannot be overstated. Every victory on the strip translated to increased prestige on the street. The phrase "Win on Sunday, Sell on Monday" was never more apt than with the Barracuda. Its NHRA triumphs not only boosted sales but also inspired a generation of gearheads to push the limits of what was possible with their own vehicles.

Moreover, the technological advancements made for NHRA competition often found their way into production models. Improved engine designs, stronger drivetrains, and more effective cooling systems all trickled down to the Barracudas that customers could buy at their local Plymouth dealerships. This direct link between racing and production not only improved the breed but also gave Barracuda owners a tangible connection to the glory of the drag strip.

As the NHRA Super Stock and Pro Stock classes evolved, so did the Barracuda. Each season brought new challenges and innovations, with Plymouth consistently at the forefront of development. The Barracuda's adaptability and the dedication of its engineering team ensured that it remained competitive even as rival manufacturers stepped up their game.

In the annals of NHRA history, the Plymouth Barracuda stands as a giant. Its success in the Super Stock and Pro Stock classes not only brought glory to the brand but also pushed the entire sport of drag racing forward. The lessons learned on the strip, the records set, and the memories created continue to resonate with enthusiasts today, ensuring that the Barracuda's legacy in NHRA competition will never be forgotten.

Section 7.5: International Racing Endeavors

The Plymouth Barracuda's influence extended far beyond American shores, making its mark in international racing events and captivating audiences worldwide. As the muscle car era reached its zenith, Plymouth recognized the importance of global recognition and set its sights on conquering racetracks across the globe.

The Barracuda's foray into international racing required careful consideration of varying regulations and track conditions. Engineers at Plymouth worked tirelessly to adapt the vehicle to meet different racing standards while maintaining its core performance characteristics. These modifications often included adjustments to engine displacement, fuel systems, and aerodynamics to comply with local rules and optimize performance for diverse racing environments.

One of the Barracuda's most notable international achievements was its participation in the grueling 24 Hours of Spa-Francorchamps in Belgium. The endurance race, known for its challenging track and unpredictable weather, provided an ideal stage for the Barracuda to showcase its durability and performance. In 1969, a specially prepared Barracuda, piloted by a team of skilled drivers, not only completed the punishing race but also secured a respectable finish, turning heads in the European racing community.

The Australian Touring Car Championship also saw the Barracuda make waves Down Under. Modified to meet local racing specifications, the Barracuda competed against a field of familiar muscle cars and unique Australian models. Its powerful engine and nimble handling characteristics translated well to the fast, flowing circuits of Australia, earning it a devoted following among Aussie racing fans.

In Japan, the Barracuda found success in exhibition races and time attack events. Its distinctive American muscle car styling and thunderous V8 roar made it a crowd favorite at tracks like Fuji Speedway and Suzuka Circuit. These appearances not only boosted the Barracuda's international racing credentials but also helped expand Plymouth's market presence in the growing Japanese automotive enthusiast scene.

The Barracuda's international racing endeavors had a profound impact on its global appeal. Success on foreign tracks translated to increased interest from international buyers, helping to establish the Barracuda as a genuinely global performance icon. Automotive journalists from Europe to Asia praised the car's versatility and competitiveness, often comparing it favorably to established international racing marques.

When pitted against its international competitors on the racing circuit, the Barracuda held its own admirably. While European sports cars often had the advantage in handling on twisty circuits, the Barracuda's raw power and straight-line speed made it a formidable opponent on faster tracks. Its battles with Ford Mustangs, Chevrolet Camaros, and even Porsche 911s in various international events became the stuff of legend, further cementing its reputation as a world-class performance machine.

The lessons learned from international racing proved invaluable for Plymouth. The diverse challenges presented by different tracks, regulations, and competitors drove continuous improvement and innovation. These experiences influenced not only future racing programs but also the development of production Barracudas, as engineers incorporated insights gained from global competition into subsequent model years.

Ultimately, the Barracuda's international racing endeavors played a crucial role in elevating its status from an American muscle car to a globally recognized performance icon. The success and exposure gained on the world stage contributed significantly to the Barracuda's enduring legacy, ensuring its place in the pantheon of great racing cars that transcended national boundaries.

Section 7.6: Factory Support and Race-Spec Models

Plymouth's commitment to racing success with the Barracuda went far beyond simply providing cars to talented drivers. The company established a robust factory racing program that played a crucial role in developing and supporting the Barracuda's competitive edge. This dedicated effort saw the creation of special race-only Barracuda models, purpose-built to dominate on the track and strip.

The factory racing program was a hotbed of innovation, where engineers and designers worked tirelessly to push the boundaries of what the Barracuda could achieve. These specialists were given the freedom to experiment with cutting-edge technologies and materials, often developing solutions that were years ahead of their time. Their work resulted in a series of race-spec Barracudas that were lighter, more powerful, and more aerodynamic than their street-legal counterparts.

One of the most notable outcomes of this program was the development of the legendary 'Super Stock' Barracudas. These purpose-built drag racers featured lightweight body panels, stripped-down interiors, and highly modified engines capable of producing astonishing amounts of horsepower. The factory also produced a limited number of 'Trans-Am' spec Barracudas, specially prepared for the rigors of road racing with enhanced suspension, braking, and aerodynamic packages.

Unleashing the Beast: A Comprehensive History of the Barracuda

The role of Plymouth's engineers and designers in this racing development cannot be overstated. These professionals worked closely with drivers and race teams, constantly refining and improving the Barracuda based on real-world racing feedback. This iterative process led to rapid advancements in areas such as engine tuning, weight reduction techniques, and aerodynamics. The lessons learned on the track often found their way into the design and engineering of production models, creating a virtuous cycle of innovation that benefited both racers and everyday Barracuda owners.

Indeed, many of the racing innovations developed for the Barracuda eventually trickled down to production models. Features like improved cooling systems, stronger drivetrain components, and more efficient engine designs all made their way from the track to the street. This transfer of technology not only improved the performance of road-going Barracudas but also enhanced their reliability and durability, further cementing the model's reputation for robust performance.

The impact of Plymouth's factory support on the Barracuda's racing success was profound. By providing top-tier equipment, technical expertise, and financial backing, Plymouth enabled talented drivers to showcase the Barracuda's true potential on the world stage. This support translated into numerous victories and record-breaking performances across various racing disciplines, from quarter-mile drag strips to challenging road courses.

Moreover, the factory's involvement lent credibility and prestige to the Barracuda's racing efforts. When potential buyers saw factory-backed Barracudas winning races and setting records, it reinforced the model's performance credentials and helped drive sales of both standard and high-performance variants. This symbiotic relationship between racing success and commercial appeal was a key factor in the Barracuda's enduring popularity throughout its production run.

In retrospect, Plymouth's dedication to factory support and the development of race-spec models was a masterstroke that paid dividends both on and off the track. It not only resulted in a string of impressive racing achievements but also drove technological advancements that benefited the entire Barracuda line. This commitment to racing excellence left an indelible mark on the Barracuda's legacy, ensuring its place in the pantheon of great American muscle cars.

Section 7.7: Legacy and Influence on Future Generations

The Plymouth Barracuda's racing legacy extends far beyond its active years on the track, leaving an indelible mark on motorsports and influencing generations of performance vehicles. As we reflect on the Barracuda's competitive history, it becomes clear that its impact reverberates through time, shaping the automotive landscape in ways both obvious and subtle.

The Barracuda's racing success had a profound influence on future Plymouth and Chrysler vehicles. The lessons learned on the track, in areas such as aerodynamics, engine tuning, and suspension design, found their way into production models, elevating the performance of street cars. This racing DNA can be traced through subsequent Mopar muscle cars and even into modern high-performance vehicles, such as the Dodge Challenger and Charger Hellcat models, which carry forward the spirit of their track-bred ancestors.

For collectors and enthusiasts, ex-racing Barracudas have become highly prized possessions. These battle-scarred warriors of the track, with their rich histories and unique modifications, often command astronomical prices at auctions. The rarity of genuine race-used Barracudas, combined with their historical significance, has created a niche market where provenance is king. A documented racing history can multiply a Barracuda's value many times over, with some examples fetching prices that rival those of exotic European sports cars.

Unleashing the Beast: A Comprehensive History of the Barracuda

The allure of the racing Barracuda hasn't faded with time. Modern-day events like vintage racing series and concours d'elegance often feature restored or preserved racing Barracudas, allowing new generations to experience the sights and sounds of these legendary machines in action. These retrospective events not only pay homage to the car's racing heritage but also serve to educate younger enthusiasts about the Barracuda's significant role in motorsports history.

When comparing the Barracuda's place in racing history to its contemporaries, it stands tall among giants. While cars like the Ford Mustang and Chevrolet Camaro may have had more mainstream recognition, the Barracuda carved out its own niche, particularly in drag racing and Pro Stock classes. Its success in these arenas helped establish Mopar as a force to be reckoned with in straight-line acceleration, a reputation that persists to this day.

The Barracuda's racing pedigree also contributed to the broader muscle car mythos. It exemplified the "win on Sunday, sell on Monday" philosophy that drove much of the muscle car era's development. The excitement generated by racing success translated directly into showroom traffic and street credibility, helping to cement the Barracuda's status as a true performance icon.

Perhaps most importantly, the Barracuda's racing legacy serves as a testament to the ingenuity and passion of the teams that campaigned these cars. From factory-backed efforts to privateer underdogs, the stories of determination, innovation, and sheer will to win continue to inspire. These tales of motorsport heroics keep the spirit of the racing Barracuda alive, ensuring that its influence will be felt for generations to come.

As we look to the future, the impact of the Barracuda's racing history continues to evolve. With the rising interest in electric vehicles and alternative powertrains, one might wonder how the lessons learned from the Barracuda's racing program might inform the development of future high-performance vehicles. The emphasis on lightweight construction, aerodynamic efficiency, and power delivery that defined the Barracuda's racing success remains relevant, even as the means of propulsion change.

Ultimately, the Plymouth Barracuda's racing legacy extends beyond the victories it achieved and the records it set. It's about the enduring spirit of competition, the pursuit of engineering excellence, and the passion it ignited in fans and enthusiasts. This legacy ensures that long after the last checkered flag fell, the Barracuda's influence on motorsports and automotive culture continues to thrive, inspiring new generations to push the boundaries of performance and design.

Unleashing the Beast: A Comprehensive History of the Barracuda

Chapter 8: Rare and Coveted: Limited Editions and Special Models

Section 8.1: The Holy Grail - 1970-1971 Hemi 'Cuda Convertible

In the realm of muscle car collecting, few vehicles command the reverence and awe of the 1970-1971 Hemi 'Cuda Convertible. Often referred to as the "Holy Grail" of muscle cars, this rare beast represents the pinnacle of Plymouth's performance offerings and stands as a testament to the unbridled power and audacity of the muscle car era.

The rarity of the Hemi 'Cuda Convertible cannot be overstated. In 1970, Plymouth produced a mere 14 examples, while 1971 saw only 11 roll off the assembly line. This scarcity is a primary factor in the car's legendary status and astronomical value in today's collector market.

What sets the Hemi 'Cuda Convertible apart from its hardtop brethren is not just its open-air configuration, but also its unique features and specifications. Under the hood lurks the crown jewel of

Mopar engines: the 426 Hemi. This monstrous power plant, nicknamed the "elephant motor" due to its size and weight, produced an advertised 425 horsepower. However, many enthusiasts believe the actual output was closer to 500 horsepower.

The Hemi 'Cuda Convertible wasn't just about raw power; it was a showcase of Plymouth's finest performance technology. It came standard with a heavy-duty suspension, upgraded brakes, and a choice of either a four-speed manual transmission or a three-speed TorqueFlite automatic. The "shaker" hood scoop, which protruded through the hood and visibly shook with the engine's vibrations, became an iconic feature of these cars.

In terms of performance, the Hemi 'Cuda Convertible was a force to be reckoned with. It could sprint from 0 to 60 mph in just 5.8 seconds and complete the quarter-mile in 14 seconds flat. These numbers were impressive for the time and still command respect today.

The value of these rare convertibles has skyrocketed over the years. In 2014, a 1971 Hemi 'Cuda Convertible made headlines when it sold for a staggering $3.5 million at auction, setting a record for American muscle cars. Even "lesser" examples routinely fetch seven-figure sums, with prices varying based on factors such as originality, documentation, and specific options.

Several factors contribute to the Hemi 'Cuda Convertible's status as the ultimate Barracuda. First, its rarity makes it a true collector's item. With only 25 examples produced over two years, it's one of the rarest production muscle cars ever made. Second, it represents the zenith of Plymouth's performance offerings, combining the most powerful engine available with the desirable convertible body style. Third, it marks the end of an era; 1971 was the last year for both the Hemi engine and the Barracuda convertible, making these cars the last of their kind.

Moreover, the Hemi 'Cuda Convertible embodies the excess and bravado of the muscle car era. It was a car that prioritized performance above all else, even if that meant sacrificing practicality and fuel efficiency. In many ways, it represents the American automotive industry at its most daring and innovative.

The 1970-1971 Hemi 'Cuda Convertible isn't just a car; it's a rolling piece of automotive history. Its combination of rarity, performance, and historical significance has elevated it to mythical status among collectors and enthusiasts. For many, owning one of these cars isn't just about having a valuable asset; it's about possessing a piece of the American dream, a tangible connection to an era when horsepower was king and the open road beckoned with endless possibilities.

Section 8.2: AAR 'Cuda - The Trans Am Special

The AAR 'Cuda stands as a testament to Plymouth's racing ambitions and represents a unique chapter in the Barracuda's storied history. Born out of necessity to compete in the Sports Car Club of America's (SCCA) Trans American Sedan Championship, the AAR 'Cuda was a street-legal homologation special that bridged the gap between track and road.

The origins of the AAR 'Cuda can be traced back to Plymouth's partnership with Dan Gurney's All American Racers (AAR) team. To qualify for the Trans Am series, manufacturers were required to produce and sell a minimum number of street versions of their race cars. This requirement gave birth to the AAR 'Cuda, a limited-production model that would become one of the most sought-after variants of the Barracuda line.

What set the AAR 'Cuda apart were its distinctive features, chief among them the potent 340 Six Pack engine. This powerplant was a force to be reckoned with, featuring three two-barrel carburetors atop a high-performance 340 cubic inch V8. The result was an official rating of 290 horsepower, though many believe this figure to be understated. The AAR 'Cuda also sported a unique side exhaust system that not only enhanced its aggressive appearance but also produced a menacing growl that left no doubt about its performance credentials.

Visually, the AAR 'Cuda was a standout. It featured a matte-black fiberglass hood with a large functional scoop, eye-catching strobe stripes along the sides, and a rear spoiler that added both style and downforce. The car sat slightly higher in the rear, giving it a purposeful, race-ready stance that hinted at its track-bred nature.

On the racing circuit, the AAR 'Cuda had a brief but impactful career. While it faced stiff competition from the likes of Ford's Boss 302 Mustang and Chevrolet's Camaro Z/28, the AAR 'Cuda managed to secure several impressive finishes. Its most notable achievement came at the 1970 Laguna Seca round, where Dan Gurney piloted the car to a hard-fought third-place finish.

Production of the AAR 'Cuda was limited to a single model year, 1970, with only 2,724 units built. This limited run was split between various color options, with some featuring the iconic "billboard" side graphics that have become highly prized by collectors. In today's market, the AAR' Cuda is a highly sought-after collector's item. Its rarity, combined with its rich racing heritage and distinctive features, has driven values to impressive heights. Well-preserved or expertly restored examples can command prices well into six figures, with particularly exceptional specimens fetching even more at auction.

The AAR 'Cuda represents more than just a high-performance variant of the Barracuda; it embodies a pivotal moment in American motorsport history. It stands as a tangible link to an era when manufacturers invested significant resources in racing programs, creating street-legal versions of their track cars for enthusiasts to enjoy. For many collectors and Mopar enthusiasts, owning an AAR 'Cuda is akin to possessing a piece of automotive racing royalty, a reminder of Plymouth's daring foray into the highly competitive world of Trans Am racing.

Section 8.3: The Rapid Transit System Caravan Cars

The Rapid Transit System (RTS) program was a unique marketing initiative launched by Plymouth in the early 1970s, designed to showcase the brand's high-performance vehicles and connect with enthusiasts across the country.

At the heart of this program were the RTS Caravan Cars, a collection of specially modified Plymouth vehicles that toured the United States, making appearances at dealerships, car shows, and other events. Among these eye-catching promotional vehicles were several custom Barracudas that left an indelible mark on the muscle car community.

The RTS Caravan 'Cudas were far from ordinary production models. These vehicles were extensively modified to create maximum visual impact and demonstrate the potential of Plymouth's performance offerings. Each Barracuda in the caravan received a unique paint job, often featuring bold colors, eye-catching graphics, and custom airbrush work that set them apart from anything else on the road. The exteriors were further enhanced with body modifications, including hood scoops, spoilers, and custom grilles, creating a look that was both aggressive and futuristic.

Under the hood, the RTS Caravan Cudas were just as impressive. Many were equipped with high-performance engine options, including the mighty 426 Hemi and the 440 Six Pack. These power plants were often dressed up with chrome accessories and custom touches to make them as visually appealing as they were powerful. The interiors of these special Barracudas were equally customized, featuring unique upholstery, gauges, and other bespoke touches that reflected the car's overall theme.

One of the most famous RTS Caravan 'Cudas was the "Mod Top fish," a 1970 model that combined the flower-power aesthetic of the Mod Top option with a hand-painted psychedelic fish mural on its sides. This car epitomized the bold, attention-grabbing nature of the

RTS program and became an instant hit with crowds wherever it appeared.

The impact of the RTS Caravan Cars on Plymouth's marketing efforts and public perception cannot be overstated. These rolling showcases of automotive artistry and performance captured the imagination of car enthusiasts across the country, helping to cement the Barracuda's reputation as one of the most exciting muscle cars of its era. The program enabled Plymouth to directly engage with potential customers, providing them with a tangible experience of the brand's performance capabilities and design prowess.

Today, the RTS Caravan 'Cudas are among the rarest and most valuable Barracudas in existence. Many of these cars were eventually returned to stock configuration and sold off, while others were unfortunately lost to time. The few that have survived in their original RTS form are highly prized by collectors and serve as fascinating time capsules of a unique moment in automotive marketing history.

The value of these promotional vehicles is difficult to quantify, as they rarely come up for sale. When they do, they often command premium prices far beyond those of even the rarest production Barracudas. Their combination of rarity, historical significance, and sheer visual impact makes them highly sought after by serious muscle car collectors and Plymouth enthusiasts.

The legacy of the RTS Caravan Cars lives on in the memories of those who saw them in person and in the annals of muscle car history. These special Barracudas represent a time when automakers were willing to push the boundaries of creativity and showmanship to capture the public's attention, resulting in some of the most unique and memorable muscle cars ever created.

Section 8.4: The Mod Top 'Cuda

The Mod Top 'Cuda stands as one of the most visually distinctive and era-defining versions of the Plymouth Barracuda. Introduced in

1969, the Mod Top option was Plymouth's bold answer to the vibrant, psychedelic aesthetics of the late 1960s. This unique feature offered buyers the opportunity to outfit their Barracuda with a vinyl top adorned in eye-catching floral patterns, perfectly capturing the spirit of the flower power generation.

The origins of the Mod Top option can be traced back to Plymouth's desire to appeal to younger, more fashion-conscious buyers. In an era where self-expression and individuality were paramount, the Mod Top 'Cuda provided a canvas for personal style right off the showroom floor. The idea was revolutionary: instead of the conventional solid-color vinyl tops, why not offer something that truly stood out?

Plymouth didn't hold back when it came to patterns and color combinations. The most famous Mod Top design featured a psychedelic floral print in shades of green, yellow, and white on a blue background. This wasn't the only option, however. Buyers could also choose from other floral patterns in different color schemes, including a yellow and orange combo that was particularly striking when paired with Vitamin C Orange paint.

What truly sets the Mod Top 'Cuda apart is its rarity. Despite the attention-grabbing nature of the option, it didn't prove to be a massive hit with buyers. As a result, production numbers were extremely low. Exact figures are hard to come by, but it's estimated that fewer than 1,000 Mod Top 'Cudas were ever produced across all Barracuda body styles. This scarcity has only added to their allure among collectors today.

When the Mod Top 'Cuda was first introduced, reception among buyers and critics was mixed. Some praised Plymouth for its daring and innovative approach to personalization, seeing the Mod Top as a perfect reflection of the times. Others found it gaudy or gimmicky, preferring the more traditional muscle car aesthetics. Regardless of individual opinions, there's no denying that the Mod Top 'Cuda made a statement wherever it went.

In today's collector market, the Mod Top 'Cuda is highly sought after for its uniqueness and historical significance. These cars serve as rolling time capsules, perfectly encapsulating the aesthetic and cultural zeitgeist of the late 1960s. However, their rarity and distinctive features present unique challenges for restoration.

Finding an original Mod Top in good condition is extremely difficult, and reproducing the intricate floral patterns with accuracy is a complex and expensive process. As a result, properly restored Mod Top 'Cudas command a significant premium in the classic car market. For many collectors, though, the opportunity to own such a unique piece of automotive and cultural history is well worth the investment.

The Mod Top 'Cuda represents more than just a quirky option package. It stands as a testament to Plymouth's willingness to take risks and push the boundaries of automotive design. In an era where their performance and aggressive styling often defined muscle cars, the Mod Top 'Cuda dared to be different, proving that even the most potent muscle cars could have a softer, more playful side. Today, these rare Barracudas serve as colorful reminders of a time when flower power met horsepower on the American roadways.

Section 8.5: The Cunningham's Cuda Race Cars

The intersection of American muscle and European racing pedigree found its embodiment in the Cunningham 'Cuda race cars. These unique vehicles represent a fascinating chapter in the Plymouth Barracuda's history, blending raw power with sophisticated racing engineering.

Briggs Cunningham, a name synonymous with American racing success in the mid-20th century, was a wealthy entrepreneur and sportsman with a passion for automobiles. His racing team had already made waves at Le Mans with Corvettes in the 1960s, but for the 1970 season, Cunningham set his sights on the Trans Am series, choosing the Plymouth Barracuda as his weapon of choice.

Unleashing the Beast: A Comprehensive History of the Barracuda

The Cunningham 'Cudas were far from standard production models. Built to Trans Am specifications, these race cars started life as 1970 Plymouth Barracuda hardtops but underwent extensive modifications. The heart of each Cunningham 'Cuda was a specially prepared 340 cubic inch V8 engine, pushing the limits of the Trans Am's 5-liter (305 cubic inch) restriction.

These engines were equipped with high-performance components, including forged pistons, a high-lift camshaft, and a tuned exhaust system, which enabled them to produce approximately 450 horsepower, a substantial increase over the stock 340's output.

The cars' bodywork was extensively modified to improve aerodynamics and reduce weight. Fiberglass hoods, fenders, and deck lids replaced the stock steel components, while the interior was stripped bare and fitted with a full roll cage, racing seat, and other safety equipment required for competition. The suspension was overhauled entirely with heavy-duty components, and large disc brakes were installed at all four corners to provide the stopping power needed on the track.

Painted in the distinctive blue and white livery of the Cunningham team, these 'Cudas made their racing debut at the 1970 Trans Am season opener at Laguna Seca. Despite facing stiff competition from factory-backed teams from Ford, Chevrolet, and AMC, the Cunningham 'Cudas proved to be formidable contenders. They achieved several top-ten finishes throughout the season, with their best result being a fourth-place finish at Road America.

The Cunningham 'Cudas' racing career was relatively short-lived, competing primarily in the 1970 and 1971 Trans Am seasons. However, their impact on the Barracuda's racing legacy was significant. They demonstrated that the Barracuda platform, when properly developed, could compete at the highest levels of American road racing.

Today, the Cunningham 'Cudas are among the most valuable and sought-after Barracudas in existence. Of the four cars initially built for the team, only three are known to have survived. These survivors have become prized possessions of collectors and museums, representing a unique blend of American muscle car heritage and professional racing pedigree.

The value of these historic racers is challenging to quantify, as they rarely change hands in public sales. However, given their rarity, racing history, and connection to both the Cunningham name and the golden era of Trans Am racing, it's safe to say that they would command multi-million dollar price tags if ever offered for sale.

The Cunningham 'Cudas serve as a testament to the Plymouth Barracuda's versatility and potential. They remind us that the Barracuda was more than just a street machine - it was a platform capable of competing at the highest levels of motorsport when placed in the right hands. These race cars represent a fascinating "what if" scenario, hinting at the racing success Plymouth might have achieved had it committed to a full factory racing program for the Barracuda.

Section 8.6: The 1970 'Cuda Gran Coupe

The 1970 Cuda Gran Coupe stands as a testament to Plymouth's desire to blend luxury with performance, creating a unique offering in the muscle car market. Conceived as a high-end variant of the already popular 'Cuda, the Gran Coupe package was designed to appeal to buyers who wanted the thrill of a powerful muscle car without sacrificing comfort and sophistication.

At its core, the Gran Coupe concept aimed to elevate the 'Cuda's status, positioning it as a more refined alternative to its muscle car brethren. This special package featured a host of luxury appointments that set it apart from the standard 'Cuda models. The exterior boasted unique badging, including distinctive "Gran Coupe" emblems, alerting onlookers to its special status. Chrome accents were applied more liberally, giving the car a more upscale appearance.

Inside, the Gran Coupe truly shone. High-quality vinyl or optional leather upholstery draped the seats, offering a level of comfort rarely seen in muscle cars of the era. The dashboard featured woodgrain accents, lending an air of sophistication to the interior. Power windows, often considered a luxury in 1970, came standard on the Gran Coupe, as did an upgraded sound system. Air conditioning, while optional, was a popular choice among Gran Coupe buyers, further emphasizing the model's focus on comfort.

Despite its luxury leanings, the Gran Coupe didn't skimp on performance. Buyers could choose from a range of potent engine options, including the robust 383 cubic-inch V8, the powerful 440 cubic-inch V8, and even the legendary 426 Hemi. This variety allowed customers to tailor their Gran Coupe's performance to their liking, from comfortable cruiser to blistering quarter-mile monster.

The reception of the 'Cuda Gran Coupe was generally positive. Automotive journalists praised its attempt to bridge the gap between luxury and performance, noting that it offered a unique proposition in the muscle car market. Many buyers appreciated the option to have their cake and eat it too, enjoying the prestige and comfort of a luxury car while also experiencing the heart-pounding performance of an aggressive muscle car.

However, sales figures for the Gran Coupe package were relatively modest. The added cost of the package, combined with the general decline in muscle car sales as the 1970s progressed, limited its appeal. Exact production numbers are difficult to obtain, as Plymouth didn't separate Gran Coupe production from overall 'Cuda numbers. However, it's estimated that only a small percentage of 1970s Cudas were equipped with this package.

Today, the 1970 Cuda Gran Coupe is a rare and sought-after model among collectors. Its unique blend of luxury and performance, combined with its limited production, makes it a valuable addition to any muscle car collection. Pristine examples, especially those equipped with the more powerful engine options, can command significant premiums at auction.

The Gran Coupe represents a fascinating chapter in the Barracuda's history, showcasing Plymouth's attempt to diversify the appeal of its flagship muscle car. While it may not have been a sales success in its time, the Gran Coupe has since become a cherished rarity, emblematic of an era when American automakers weren't afraid to experiment with the muscle car formula.

Section 8.7: Other Notable Special Editions

While the Hemi 'Cuda and AAR 'Cuda often steal the spotlight, the Plymouth Barracuda's history is peppered with other noteworthy special editions that deserve recognition. These models, though perhaps not as famous as their counterparts, played crucial roles in shaping the Barracuda's legacy and continue to captivate collectors and enthusiasts alike.

The 1970 Cuda 440+6 stands as a testament to Plymouth's commitment to raw power. This model featured the formidable 440 cubic inch V8 engine with three two-barrel carburetors, affectionately known as the "Six Pack." Producing an impressive 390 horsepower, the 440+6 offered Hemi-like performance at a more affordable price point. Its rarity, combined with its blistering performance, makes it a highly sought-after model among collectors today.

The 1971 Barracuda Convertible holds a special place in Barracuda history as it marked the final year for the convertible body style. With tightening safety regulations on the horizon, Plymouth decided to discontinue the convertible option after 1971. This decision, coupled with relatively low production numbers, especially for high-performance variants, has made these last-of-their-kind convertibles particularly valuable in today's market.

One of the most visually striking special editions was the 1970-71 'Cuda with billboard graphics. These bold, eye-catching side stripes prominently displayed the engine size, leaving no doubt about the car's performance potential. Available in various colors and with different engine options, the billboard 'Cudas have become iconic representations of the muscle car era's exuberant styling.

The 1969 'Cuda 440 deserves mention as it marked the first time the big-block 440 cubic inch engine was offered in the Barracuda. This move signaled Plymouth's serious intentions in the muscle car wars, providing enthusiasts with a potent performance option that could hold its own against competitors from Ford and Chevrolet. The '69 Cuda 440's combination of the A-body platform and the powerful 440 engine created a uniquely balanced muscle car that remains highly regarded.

Lastly, we can't overlook the 1968 Formula S package. While not as powerful as later performance models, the Formula S played a crucial role in establishing the Barracuda's performance credentials. Offering upgraded suspension, wider tires, and a choice of small-block V8 engines, the Formula S package transformed the Barracuda from a sporty compact into a genuine performer. It set the stage for the high-horsepower models that would follow and holds a special place in Barracuda history as a harbinger of things to come.

Each of these special editions contributed to the rich tapestry of the Plymouth Barracuda's history. Whether it was through groundbreaking performance, unique styling elements, or simply marking a significant moment in the model's evolution, these cars helped cement the Barracuda's place in muscle car lore. Today, they serve as prized possessions for collectors and vivid reminders of an era when American automakers pushed the boundaries of performance and style.

Unleashing the Beast: A Comprehensive History of the Barracuda

Chapter 9: The Barracuda's Place in the Muscle Car Era

Section 9.1: Defining the Muscle Car Era

The muscle car era, a golden age in American automotive history, was a period of unprecedented power, performance, and style that captivated the nation from the early 1960s to the mid-1970s. This era was characterized by the production of high-performance vehicles that combined powerful engines with affordable, mid-size car bodies, creating a perfect storm of speed and accessibility for the average consumer.

The timeline of the muscle car era can be traced back to 1964, with the introduction of the Pontiac GTO, widely considered the first solid muscle car. This revolutionary vehicle set the stage for a decade of intense competition among American automakers, each vying to produce the fastest, most powerful cars on the market. The era reached its peak in the late 1960s and early 1970s, with iconic models like the Chevrolet Chevelle SS, Ford Mustang Boss 429, and, of course, the Plymouth Barracuda, dominating the streets and drag strips across the country.

Unleashing the Beast: A Comprehensive History of the Barracuda

Several cultural and economic factors drove the muscle car trend. The post-World War II economic boom had created a generation of young, affluent consumers eager for excitement and self-expression. The emerging youth culture of the 1960s, with its emphasis on rebellion and individuality, found a perfect outlet in these powerful, eye-catching vehicles. Additionally, the relatively low price of gasoline and the absence of strict emissions regulations allowed automakers to focus on performance above all else.

To be considered an actual muscle car, a vehicle needed to meet specific criteria. First and foremost was a powerful engine, typically a V8 with high horsepower output. These engines were often derived from or inspired by racing technology, bringing track performance to the street. Secondly, muscle cars were generally mid-size, two-door vehicles built on existing passenger car platforms, making them more affordable and practical than dedicated sports cars. Lastly, they featured aggressive styling cues, such as hood scoops, bold graphics, and wide tires, that clearly communicated their performance potential.

The major players in the muscle car market were the "Big Three" American automakers: General Motors, Ford, and Chrysler. Each company had multiple divisions producing their own variants of muscle cars. GM offered the Pontiac GTO, Chevrolet Chevelle SS, and Oldsmobile 442, among others. Ford countered with the Mustang, Torino, and Mercury Cougar. Chrysler's arsenal included the Plymouth Road Runner, Dodge Charger, and, of course, the Plymouth Barracuda. These companies engaged in a horsepower war, constantly one-upping each other with more powerful engines and outrageous performance claims.

The public's perception and reception of muscle cars was overwhelmingly positive during the peak of the era. These vehicles represented freedom, power, and the American dream for many. Young buyers flocked to dealerships, eager to get their hands on the latest and most excellent models. Muscle cars became status symbols and objects of desire, featured prominently in movies, television shows, and popular music of the time. They were not just modes of transportation, but cultural icons that embodied the spirit of an era.

However, as the 1970s progressed, public opinion began to shift. Rising insurance rates, stricter safety and emissions regulations, and the 1973 oil crisis all contributed to a decline in the popularity and feasibility of muscle cars. The era that had burned so brightly was beginning to fade, but not before leaving an indelible mark on automotive history and American culture.

The muscle car era, with its emphasis on raw power and bold styling, set the stage for the Plymouth Barracuda to make its mark. As we'll explore in the following sections, the Barracuda would carve out its own unique position within this competitive and dynamic landscape, contributing significantly to the muscle car legacy that continues to captivate enthusiasts to this day.

Section 9.2: The Barracuda's Unique Position

The Plymouth Barracuda occupied a distinctive place in the muscle car landscape, evolving from a pioneering pony car to a formidable muscle machine. When first introduced in 1964, the Barracuda was actually ahead of its time, beating the Ford Mustang to market by two weeks. This early debut positioned the Barracuda as a trendsetter, though it would soon find itself competing in a rapidly crowding field.

The Barracuda's journey from pony car to muscle car was a gradual but deliberate transformation. Initially built on the compact Valiant platform, the first-generation Barracuda was more of a sporty compact than a muscle car. However, with each subsequent redesign, Plymouth pushed the Barracuda further into muscle car territory. By the time the third generation rolled out in 1970, the Barracuda had shed its compact car roots entirely, embracing a purpose-built performance platform shared with the Dodge Challenger.

What truly set the Barracuda apart from its competitors was its distinctive styling. While many muscle cars of the era adopted a similar long-hood, short-deck profile, the Barracuda - especially in its final generation - stood out with its unique design elements. The wraparound rear window of the first generation was a bold styling choice that set it apart from anything else on the road. The third-generation 'Cuda, with its aggressive grille, deeply inset headlights, and muscular fender flares, exuded a menacing presence that few other muscle cars could match.

Performance-wise, the Barracuda offered a wide range of capabilities across different models. From the base six-cylinder engines to the fire-breathing Hemi V8s, there was a Barracuda for every type of driver. The top-tier models, particularly the Hemi 'Cuda, could go toe-to-toe with the most potent muscle cars of the era, while more modest versions provided an accessible entry point for buyers wanting muscle car style without the high-performance price tag.

Perhaps most interestingly, the Barracuda appealed to a diverse range of demographics. Its stylish design and performance credentials made it popular among young, performance-oriented buyers. At the same time, its Plymouth badging and more luxurious Gran Coupe variants attracted older, more conservative customers who might have been put off by some of the more outrageous muscle cars of the time. This broad appeal helped the Barracuda carve out its own niche in a crowded market.

The Barracuda's unique position in the muscle car era was defined by its early entry into the market, its evolutionary journey from pony car to muscle car, its distinctive styling, varied performance offerings, and broad demographic appeal. These factors combined to create a vehicle that was more than just another entry in the muscle car wars - it was a unique player that helped shape the era itself.

Section 9.3: Iconic Barracuda Models in the Muscle Car Pantheon

The Plymouth Barracuda's journey through the muscle car era was marked by several standout models that cemented its place in automotive history. These iconic variants not only showcased the Barracuda's evolution but also represented the pinnacle of Mopar muscle.

At the top of the Barracuda hierarchy stood the 1970-1971 Hemi 'Cuda. This beast of a machine was the ultimate expression of muscle car excess, powered by the legendary 426 cubic inch Hemi V8 engine. Producing a conservatively rated 425 horsepower, the Hemi 'Cuda was a force to be reckoned with on both the street and the strip. Its rarity and performance potential have made it one of the most sought-after muscle cars of all time, with pristine examples fetching millions at auction.

The 1970 AAR 'Cuda holds a special place in Barracuda lore. Built to homologate Plymouth's Trans Am racing efforts, this limited-production model featured a unique 340 cubic inch "Six Pack" engine with three two-barrel carburetors. With its distinctive side-exit exhaust and matte black hood, the AAR 'Cuda was a street-legal race car that embodied the spirit of the era.

For those seeking a blend of luxury and raw power, the 1971 440+6 Barracuda delivered in spades. Equipped with the massive 440 cubic inch V8 and three two-barrel carburetors (hence the "+6" designation), this model offered near-Hemi levels of performance with a smoother power delivery. It represented the zenith of big-block muscle before increasing emissions regulations began to stifle performance.

Perhaps the most prized of all Barracudas is the 1970 'Cuda Convertible, especially when equipped with the Hemi engine. With only a handful ever produced, these drop-top muscle cars represent the rarest of the rare. Their scarcity and desirability have made them some of the most valuable muscle cars ever built, with prices often exceeding those of exotic European sports cars.

Beyond these headline-grabbing models, the Barracuda lineup included other noteworthy variants that contributed to its muscle car credibility. The 383 and 440 four-barrel models offered impressive performance at a more attainable price point, while the small-block 340 and 360 cars provided a nimble alternative to their big-block brethren.

Special edition models like the Rapid Transit System (RTS) 'Cudas, dealer-modified cars, and even factory-built drag racing packages further expanded the Barracuda's reach within the performance community. These variants showcased the model's versatility and appeal across different segments of the muscle car market.

Each of these iconic Barracuda models played a crucial role in establishing the car's reputation during the muscle car era.

From the earth-shaking power of the Hemi to the balanced performance of the AAR, these variants demonstrated Plymouth's commitment to pushing the boundaries of what was possible in a production car. Today, they stand as enduring symbols of an era when performance was king and the Barracuda ruled its own corner of the muscle car ocean.

Section 9.4: Performance Benchmarks

The Plymouth Barracuda's performance capabilities were a crucial factor in establishing its reputation during the muscle car era. To truly appreciate the Barracuda's prowess, we must examine its performance benchmarks and how they compare to its contemporaries.

When it came to straight-line acceleration, the top-tier Barracuda models were formidable contenders. The 1970-1971 Hemi 'Cuda, equipped with the legendary 426 cubic inch Hemi V8, was capable of sprinting from 0 to 60 mph in approximately 5.6 seconds. This was an impressive feat for the time, placing it among the fastest production cars of its era. In quarter-mile runs, the Hemi 'Cuda consistently clocked times in the low 13-second range, with trap speeds exceeding 105 mph.

The 440+6 Barracuda, while not as quick as its Hemi-powered sibling, was no slouch either. With its 440 cubic inch V8 topped by three two-barrel carburetors, it could reach 60 mph from a standstill in about 5.8 seconds and complete the quarter-mile in the mid-13-second range.

These performance figures put the Barracuda in direct competition with other muscle car titans of the era. The Chevrolet Chevelle SS 454 and Ford Mustang Boss 429 were among its closest rivals in terms of acceleration. The Barracuda often traded blows with these competitors in magazine comparison tests, sometimes coming out on top, other times falling just short, but constantly proving itself a worthy adversary.

Contemporary automotive publications were quick to recognize the Barracuda's performance potential. Car and Driver, in a 1970 road test of the Hemi 'Cuda, praised its "astonishing acceleration" and noted that it was "one of the fastest cars ever to come out of Detroit." Motor Trend similarly lauded the car's straight-line performance, while also commending its improved handling compared to earlier models.

However, magazine tests only tell part of the story. Real-world experiences from Barracuda owners often painted an even more impressive picture. Many owners reported that with proper tuning and ideal conditions, their cars could outperform the published figures. Stories of Hemi 'Cudas dipping into the high 12-second range in the quarter-mile were not uncommon among enthusiasts who had honed their launching and shifting techniques.

The Barracuda's reputation among racers and tuners was equally stellar. Its robust engine options and relatively lightweight body made it a favorite for drag racers. The car's potential for modification was significant, with many tuners extracting even more power from the already potent powerplants. In Stock and Super Stock classes at drag strips across America, Barracudas were a common sight and frequent winners.

It's worth noting that while straight-line performance was the primary focus during the muscle car era, the later Barracuda models also showed improvements in handling and braking. The introduction of the E-body platform in 1970 gave the car a wider stance and better weight distribution, enhancing its overall performance envelope beyond just acceleration.

The AAR 'Cuda, in particular, demonstrated the Barracuda's all-around performance capabilities. Designed for Trans Am racing, this model featured upgraded suspension, wider tires, and a unique 340 cubic inch engine that provided a more balanced performance package. While not as brutally fast in a straight line as its big-block brethren, the AAR 'Cuda showcased the platform's potential for well-rounded performance.

In its era, the Plymouth Barracuda's performance benchmarks were truly impressive. It stood toe-to-toe with the best muscle cars of its time, often surpassing them in raw acceleration. These capabilities weren't just numbers on a spec sheet; they translated into real-world performance that solidified the Barracuda's place in the muscle car hierarchy. For many enthusiasts, the Barracuda represented the pinnacle of American performance engineering in the early 1970s, a reputation that continues to fuel its legendary status today.

Section 9.5: The Barracuda's Influence on Muscle Car Design

The Plymouth Barracuda wasn't just another player in the muscle car era; it was a trendsetter that left an indelible mark on automotive design. Throughout its production run, the Barracuda introduced several innovative styling elements that would go on to influence not only its direct competitors but the broader automotive industry as well.

One of the most notable contributions of the Barracuda to muscle car design was its distinctive fastback roofline, particularly in its first generation. The massive rear window, which wrapped around to the sides, was a bold design choice that set the Barracuda apart from its contemporaries. This unique feature not only gave the car a sleek, aerodynamic appearance but also improved rear visibility – a practical benefit that didn't go unnoticed by drivers and critics alike.

The impact of the Barracuda's design on its competitors was significant. As the pony car wars heated up, other manufacturers took note of the Barracuda's sleek profile and began incorporating similar design elements into their own models. The long hood, short deck proportions that became synonymous with muscle cars were perfectly exemplified by the Barracuda, especially in its later generations.

In terms of pushing performance boundaries, the Barracuda played a crucial role. Its ability to house Chrysler's most significant engines, including the legendary 426 Hemi, in a relatively compact body forced other manufacturers to up their game. The Barracuda proved that a smaller car could pack a serious punch, leading to an arms race of sorts among muscle car manufacturers to squeeze more power into their offerings.

The influence of the Barracuda extended beyond its immediate rivals and had a lasting impact on future Plymouth and Chrysler models. The aggressive stance, bold graphics, and high-impact color options that became hallmarks of the Barracuda, particularly in its 'Cuda incarnation, set a template for performance-oriented Chrysler vehicles for years to come. Elements of the Barracuda's design language can be seen in later models, such as the Dodge Challenger, and even in modern interpretations of muscle cars.

The most enduring aspect of the Barracuda's design legacy is its timeless appeal. The clean lines, perfect proportions, and aggressive stance of the third-generation Barracuda, in particular, continue to be celebrated by enthusiasts and designers alike. Its influence can be seen in modern muscle cars and performance vehicles, which often pay homage to the classic designs of the '60s and '70s.

The Barracuda's contributions to muscle car design weren't limited to aesthetics. Its focus on performance-oriented features, such as the Shaker hood scoop and rear window louvers, became iconic elements that were widely imitated. These functional design elements not only enhanced performance but also became visual shorthand for high-performance capabilities.

In conclusion, the Plymouth Barracuda's influence on muscle car design was far-reaching and long-lasting. From its innovative fastback design to its perfect embodiment of muscle car proportions, the Barracuda helped define the look of an era. Its legacy lives on not just in the collectible status of surviving models but in the DNA of performance cars that continue to draw inspiration from this iconic design.

Section 9.6: Challenges and Controversies

The Plymouth Barracuda's journey through the muscle car era was not without its share of obstacles and contentious issues. Like many of its high-performance counterparts, the Barracuda faced a series of challenges that would ultimately contribute to the decline of the muscle car golden age.

One of the most significant hurdles for Barracuda owners and prospective buyers was the steep rise in insurance rates. As muscle cars gained popularity, their powerful engines led to an increase in speed-related accidents, prompting insurance companies to view these vehicles as high-risk investments.

Consequently, premiums for muscle cars, including the Barracuda, skyrocketed. This financial burden made ownership increasingly difficult, especially for younger drivers who were a key demographic for these performance machines. The inflated insurance costs effectively priced many potential buyers out of the market, leading to a decline in sales and enthusiasm for the Barracuda and its muscular brethren.

Safety concerns and subsequent regulations also played a crucial role in shaping the Barracuda's fate. As public awareness of automotive safety grew, so did criticism of the muscle car's emphasis on raw power over occupant protection. The National Highway Traffic Safety Administration (NHTSA), established in 1970, began implementing stricter safety standards for all vehicles. These new regulations often required design changes that added weight and complexity to cars, ultimately affecting their performance. For the Barracuda, this meant compromises in design and engineering that slowly chipped away at its pure muscle car ethos.

The 1973 oil crisis dealt another significant blow to the Barracuda and the entire muscle car segment. As fuel prices soared and gasoline shortages became commonplace, the thirsty V8 engines that were the hallmark of muscle cars suddenly became a liability. Consumer preferences shifted rapidly towards more fuel-efficient vehicles, leaving gas-guzzling performance cars like the Barracuda out of favor. This global event accelerated the decline of the muscle car era and forced manufacturers, including Plymouth, to rethink their product strategies.

Interestingly, the Barracuda also faced challenges from within its own family. Competition within Chrysler's lineup, particularly from the Dodge Challenger, created a complex market dynamic. While the two cars shared many components, their slight differences in styling and marketing often led to internal competition for resources and sales. This sibling rivalry sometimes resulted in compromises that affected the Barracuda's development and market positioning.

As the 1970s progressed, public perception of muscle cars began to shift. The once-celebrated icons of American performance began to be seen as relics of a bygone era, out of touch with the new realities of fuel efficiency and environmental consciousness. The Barracuda, despite its passionate fan base, was not immune to this changing tide of public opinion. Its image as a gas-guzzling, high-powered machine became increasingly at odds with the evolving automotive landscape.

These challenges and controversies collectively contributed to the Barracuda's eventual demise in 1974. However, they also serve to highlight the complex interplay of social, economic, and regulatory factors that shaped the muscle car era. The Barracuda's struggle against these forces underscores its significance as a symbol of a unique period in automotive history, a time when performance reigned supreme, but was ultimately tempered by the realities of a changing world.

The controversies and challenges faced by the Barracuda during this period have become an integral part of its lore, adding depth to its story and contributing to its enduring appeal among enthusiasts. These obstacles have only served to enhance the Barracuda's status as a true muscle car icon, representing both the highs and lows of one of the most exciting eras in American automotive history.

Section 9.7: The Barracuda's Enduring Appeal

The Plymouth Barracuda may have ceased production decades ago, but its allure continues to captivate automotive enthusiasts and collectors alike. This enduring appeal is a testament to the car's iconic status and its significant place in muscle car history. One of the most tangible indicators of the Barracuda's lasting impact is its collectibility and value appreciation over time.

Rare models, particularly the 1970-1971 Hemi 'Cudas, have become some of the most sought-after muscle cars in the world. In recent years, pristine examples have fetched millions of dollars at auction, with the rarest convertible models commanding astronomical sums. This dramatic increase in value reflects not only the car's rarity but also its historical significance and the nostalgia it evokes for the golden age of American performance cars.

The Barracuda's presence in pop culture and media has played a crucial role in maintaining its iconic status. From appearing in classic films and TV shows to being featured in modern video games, the Barracuda has remained in the public eye long after it left the production line. Its distinctive styling and powerful image have made it a favorite choice for directors and producers looking to evoke the spirit of the muscle car era or to give their characters a vehicle with instant cool factor.

Enthusiast communities and clubs dedicated to the Barracuda have flourished over the years, bringing together owners, restorers, and admirers of these classic machines. These groups organize meets, share restoration tips, and work to preserve the history and legacy of the Barracuda. Their passion ensures that knowledge about these cars is passed down to new generations of enthusiasts, keeping the Barracuda's story alive and relevant.

The influence of the Barracuda on modern performance vehicles is undeniable. While the nameplate itself hasn't been revived, elements of its design and performance philosophy can be seen in contemporary muscle cars. The bold styling, aggressive stance, and focus on raw power that characterized the Barracuda have inspired modern interpretations in vehicles like the Dodge Challenger and Charger. These spiritual successors carry forward the ethos of the muscle car era, with the Barracuda serving as a key touchstone for their designers and engineers.

Today, the Plymouth Barracuda stands as a true muscle car icon. Its status has been cemented not just by its performance and styling during its production years, but by the lasting impression it has made on car culture. For many, the Barracuda represents the pinnacle of the muscle car era, a time when American automakers pushed the boundaries of performance and design with little regard for practicality or restraint.

The Barracuda's enduring appeal lies in its ability to embody the spirit of an entire era. It represents a time of automotive optimism and excess, when the roar of a powerful engine and the thrill of acceleration were celebrated without reservation. For collectors, enthusiasts, and casual admirers alike, the Barracuda continues to symbolize the freedom, power, and unbridled excitement that defined the muscle car era.

As we look back on the Barracuda's legacy, it's clear that its impact extends far beyond its relatively short production run. It remains a benchmark against which other muscle cars are measured, a dream car for many enthusiasts, and a powerful symbol of American automotive prowess. The Plymouth Barracuda may no longer prowl the streets in significant numbers, but its legend lives on, ensuring its place in the pantheon of automotive greats for generations to come.

Unleashing the Beast: A Comprehensive History of the Barracuda

Chapter 10: Competitors and Contemporaries: How the 'Cuda Stacked Up

Section 10.1: The Pony Car Corral

The muscle car era of the 1960s and '70s was a time of fierce competition, with each manufacturer vying for supremacy on the streets and in the showrooms. At the heart of this automotive arms race was the pony car segment, a category that the Plymouth Barracuda found itself squarely in the middle of. To truly appreciate how the 'Cuda stacked up, we must first examine its primary rivals in this high-octane corral.

Leading the pack was the Ford Mustang, the car that gave the pony car category its name. Introduced in 1964, the Mustang set the template that others would follow: a long hood, short deck, and a potent range of engines. The Mustang's instant success sent shockwaves through the industry, prompting a swift response from competitors. Its blend of style, performance, and affordability made it the benchmark against which all other pony cars, including the Barracuda, would be measured.

Unleashing the Beast: A Comprehensive History of the Barracuda

Hot on the Mustang's heels came the Chevrolet Camaro in 1967. General Motors' answer to the pony car challenge brought with it a range of powerful engines and aggressive styling. The Camaro quickly established itself as a formidable competitor, often going head-to-head with the Mustang in sales and performance comparisons. Its introduction raised the stakes in the pony car wars, forcing Plymouth to up its game with the Barracuda.

Alongside the Camaro, GM also introduced the Pontiac Firebird. Sharing a platform with its Chevrolet sibling, the Firebird distinguished itself with unique styling cues and Pontiac's own range of engines. The Firebird, particularly in its potent Trans Am guise, became a symbol of American muscle and a worthy adversary for the Barracuda in both performance and desirability.

Often overlooked but no less significant was the AMC Javelin. American Motors Corporation, the smallest of the American automakers, proved it could run with the big boys when it introduced the Javelin in 1968. Despite more limited resources, AMC managed to create a pony car that could hold its own in terms of style and performance, offering buyers an alternative to the offerings from the Big Three.

Last but certainly not least was the Dodge Challenger, introduced in 1970. The Challenger represented an interesting dynamic in the pony car corral: a sibling rivalry. Sharing a platform with the Plymouth Barracuda, the Challenger was Dodge's entry into the pony car market. While mechanically similar to the 'Cuda, the Challenger sported its own distinct styling and marketing approach, creating a unique in-house competition that pushed both Mopar brands to excel.

Each of these competitors brought something unique to the table, forcing the Barracuda to evolve and improve to stay competitive continually. The pony car corral was a hotbed of innovation, with each manufacturer pushing the envelope in terms of design, performance, and marketing. It was in this crucible of competition that the Barracuda would forge its identity, transforming from a compact sports car into a true muscle car icon.

As we delve deeper into how the 'Cuda measured up against these rivals, we'll see how Plymouth leveraged its strengths, addressed its weaknesses, and ultimately carved out its own niche in this crowded and competitive field. The story of the Barracuda is not just about one car, but about how it stood tall among giants in one of the most exciting eras of automotive history.

Section 10.2: Performance Face-Off

When it came to raw performance, the Plymouth Barracuda was a force to be reckoned with in the muscle car arena. The horsepower wars of the late 1960s and early 1970s saw manufacturers constantly one-upping each other, and Plymouth was no exception. The Barracuda's engine options ranged from modest six-cylinders to the legendary 426 Hemi, offering something for every level of performance enthusiast.

At the heart of the muscle car experience was quarter-mile performance, and the Barracuda held its own on the drag strip. The top-tier Hemi 'Cuda could blast through the quarter-mile in the low 13-second range, a time that could make Mustangs and Camaros nervous. Even the more common 383 and 440 engines provided stout performance, with times in the mid-13 to low-14 second range, depending on gearing and driver skill. These times were competitive with, and in some cases superior to, comparable offerings from Ford, Chevrolet, and Pontiac.

But muscle cars weren't just about straight-line speed, and the Barracuda proved it could handle curves as well. The E-body platform, shared with the Dodge Challenger, provided a stable foundation for cornering prowess. With its wide stance and well-tuned suspension, particularly in models equipped with the performance-oriented packages, the 'Cuda could carve through corners with surprising agility for its size. Contemporary road tests often praised the Barracuda's balanced handling, noting that it felt more composed in tight turns than some of its pony car rivals.

Stopping power was equally important, and Plymouth didn't skimp in this department. As horsepower figures climbed, so did the need for robust braking systems. The Barracuda offered impressive stopping power, especially when equipped with the optional four-wheel disc brakes. These brakes provided fade-resistant performance that matched or exceeded that of its competitors, a crucial factor in both street and track driving.

Safety features like improved lighting and reinforced body structures also kept pace with industry standards, ensuring the Barracuda wasn't just fast, but also relatively safe for its time. The crown jewel in the Barracuda's performance lineup was undoubtedly the Hemi 'Cuda. This special performance package transformed the already potent Barracuda into a street-legal race car. With its 426 cubic inch Hemi V8 engine, the Hemi 'Cuda was a quarter-mile monster, capable of sub-13 second passes in skilled hands.

This put it in an elite class, able to go toe-to-toe with the likes of the Chevrolet Chevelle SS 454, Ford Mustang Boss 429, and even exotic muscle like the Pontiac GTO Judge Ram Air IV. The Hemi 'Cuda wasn't just competitive; in many cases, it was the benchmark against which other muscle cars were measured.

It's worth noting that while the Hemi 'Cuda was the performance pinnacle, it was also rare and expensive. More common were the 383 and 440 engines, which still provided exhilarating performance at a more accessible price point. These variants allowed the Barracuda to compete across a broader spectrum of the muscle car market, from affordable performance to premium power.

In the realm of muscle car performance, the Plymouth Barracuda proved it could swim with the fastest fish in the sea. From drag strip dominance to competent handling and strong braking, the 'Cuda demonstrated that it wasn't just another pretty face in the pony car crowd. It was a serious performance machine capable of holding its own against the best Detroit had to offer, cementing its place in the muscle car hierarchy.

Section 10.3: Style and Design Showdown

The muscle car era was as much about visual impact as raw performance, and the Plymouth Barracuda held its own in the style stakes against its fierce competitors. This section delves into how the 'Cuda's aesthetics and design features compare to its contemporaries, exploring everything from body lines to interior amenities.

When it came to exterior aesthetics, the Barracuda underwent a dramatic transformation over its lifetime. The first-generation models, based on the Valiant platform, were distinguished by their unique fastback design and massive rear window. While this look set them apart, it didn't quite capture the aggressive stance of the Ford Mustang or the sleek lines of the Chevrolet Camaro. However, the second and third-generation Barracudas came into their own, adopting a more muscular and purposeful appearance that rivaled the best in the business.

The 1970-1974 models, in particular, showcased a design language that perfectly blended aggression with elegance. The wide, low-slung stance, pronounced fender flares, and optional shaker hood scoop gave the 'Cuda a menacing presence that could go toe-to-toe with the Pontiac Firebird's sharp edges or the Dodge Challenger's imposing bulk. The 'Cuda's distinctive "gills" on the fenders and the wraparound rear window on the hardtop models were unique styling cues that set it apart in a crowded field.

Inside the cabin, the Barracuda offered a mix of comfort and sportiness that was competitive with its rivals. While perhaps not as plush as some higher-end Firebird interiors or as driver-focused as the Camaro's cockpit, the 'Cuda's interior struck a balance that appealed to a wide range of buyers. The high-back bucket seats, wood-grain accents (in specific models), and driver-oriented dashboard layout created an environment that was both comfortable for cruising and suitable for performance driving.

Customization was a key factor in the muscle car world, and Plymouth ensured the Barracuda could be tailored to individual tastes. The range of available paint colors, including the iconic High Impact colors like Lime Light, Vitamin C, and In-Violet, allowed for eye-catching personalization that could match or exceed options from Ford or GM. Stripe packages, hood options, and wheel choices further expanded the possibilities for making a 'Cuda stand out from the crowd.

Special editions played a significant role in the pony car wars, and the Barracuda held its own in this arena. The AAR 'Cuda, created to homologate the car for Trans Am racing, offered a unique appearance package that rivaled special editions like the Camaro Z/28 or Mustang Boss 302. The ultra-rare Hemi 'Cuda convertibles, with their combination of unparalleled power and open-air cruising, created a mystique that few competitors could match.

In terms of design innovations, the Barracuda introduced several unique features. The enormous rear window of the first-generation models was a true innovation, offering unparalleled rear visibility (albeit with some practical drawbacks). The integration of the grille and bumper in later models created a clean, modern look that influenced automotive design beyond the muscle car era. The optional "Slap Stik" shifter for automatic transmissions combined ease of use with a sporty feel, a feature that stood out among its contemporaries.

While each pony car had its strengths in design and style, the Plymouth Barracuda managed to carve out a distinct identity. It evolved from somewhat unconventional beginnings to become one of the most visually striking and desirable muscle cars of its era. The 'Cuda's ability to balance aggressive performance cues with classic proportions and unexpected elegance allowed it to stand tall in comparison to its rivals, creating a lasting image that continues to captivate enthusiasts decades later.

Section 10.4: Market Position and Pricing

The Plymouth Barracuda's position in the competitive muscle car market was as dynamic as its performance on the streets. To truly understand how the 'Cuda stacked up against its rivals, we need to examine its market position and pricing strategy across various trim levels and years.

At the entry level, the base model Barracuda aimed to capture budget-conscious buyers looking for style without breaking the bank. Compared to the Ford Mustang's base offering, the 'Cuda often came in slightly higher, but remained competitive. The Chevrolet Camaro, introduced in 1967, typically undercut in its most basic form. AMC's Javelin, when it arrived in 1968, often offered the most attractive base price, reflecting the company's need to compete aggressively against the Big Three.

Moving up to mid-range trims, the Barracuda found itself in a fierce battle for the hearts and wallets of enthusiasts. These models, often featuring upgraded engines and more luxurious interiors, represented the sweet spot for many buyers. The 'Cuda's mid-range offerings were priced competitively with comparable Mustang and Camaro models, often providing more standard features for the money. Pontiac's Firebird, sharing a platform with the Camaro, typically commanded a slight premium in this segment due to its more upscale brand positioning.

At the high end of the spectrum, the Barracuda, particularly in its potent 'Cuda form, was priced to compete with the most powerful muscle cars of the era. The Hemi 'Cuda, while expensive, offered unmatched performance for the dollar. It went toe-to-toe with vehicles like the Shelby GT500, Chevrolet Camaro Z/28, and even more exotic fare like the Corvette in terms of pricing. The Dodge Challenger R/T, sharing a platform with the E-body Barracuda, was often priced similarly, creating an interesting in-house rivalry.

Sales figures tell an intriguing story of the Barracuda's market performance. While it never achieved the astronomical sales numbers of the Mustang, which benefited from being first to market and massive marketing support, the 'Cuda held its own. It consistently outsold the AMC Javelin and frequently traded places with the Pontiac Firebird in the sales rankings. However, it lagged behind the Camaro in overall sales throughout most of its run.

The target demographics for the Barracuda evolved over its lifetime. In its early years, it appealed to a slightly older, more sophisticated buyer than the Mustang, which was marketed heavily towards younger drivers. As the '60s progressed and the muscle car war heated up, Plymouth increasingly targeted performance enthusiasts, particularly with the introduction of the 'Cuda models.

This shift aligned its customer base more closely with that of the Camaro and Firebird. Interestingly, the Barracuda, especially in its later years, attracted a higher percentage of female buyers than some of its competitors. This could be attributed to its sleek design and the availability of more luxury-oriented trims alongside the high-performance variants.

The pricing and market position of the Barracuda reflected Plymouth's strategy of offering a diverse range of options to cater to different segments of the muscle car market. From affordable base models to ultra-high-performance Hemi 'Cudas, Plymouth ensured that there was a Barracuda for nearly every taste and budget. While it may not have dominated the sales charts, the Barracuda carved out a unique and respected position in one of the most competitive and exciting eras of American automotive history.

Unleashing the Beast: A Comprehensive History of the Barracuda

Section 10.5: Media and Public Perception

The Plymouth Barracuda's standing in the muscle car pantheon was not just determined by its performance on the streets and racetracks, but also by how it was perceived in the media and by the public at large. This perception played a crucial role in shaping the 'Cuda's image and legacy.

Magazine reviews of the era were a primary source of information for car enthusiasts, and the Barracuda generally fared well under the scrutiny of automotive journalists. Car and Driver praised the 1970 Hemi 'Cuda as "the best handling car of its type we've ever tested," while Motor Trend highlighted its "instant throttle response and neck-snapping acceleration." However, reviewers also noted its thirsty fuel consumption and sometimes criticized its weight, especially in later models.

In comparison, the Mustang often received accolades for its broader appeal, while the Camaro was frequently lauded for its balanced performance. The advertising battles between muscle car manufacturers were fierce and creative. Plymouth's marketing team positioned the Barracuda as a more exclusive and powerful option compared to its rivals.

Memorable ad slogans like "The way to Plymouth is marked 'Cuda" and "Cricket is a game, not a car" aimed to differentiate the 'Cuda from both its competitors and Plymouth's own economy models. In contrast, Ford's Mustang ads often emphasized its pioneering status and broad appeal, while Chevrolet positioned the Camaro as the everyman's performance car.

In popular culture, the Barracuda made its mark, but not as prominently as some of its rivals. While the Mustang starred in films like "Bullitt" and the Camaro became Bumblebee in the "Transformers" franchise, the 'Cuda had its moment, too. It featured in TV shows like "Nash Bridges" and made appearances in numerous films, albeit often in supporting roles. The distinctive sound of its Hemi engine became a cultural touchstone, recognized by gearheads in movies and TV shows even when the car itself wasn't visible.

Enthusiast communities formed around all the major muscle cars, and the Barracuda was no exception. While perhaps not as large as Mustang or Camaro clubs, 'Cuda enthusiast groups were known for their passion and dedication. These communities played a crucial role in preserving the car's legacy, organizing events, and sharing restoration tips and expertise. The exclusivity of specific 'Cuda models, particularly the limited-production Hemi versions, fostered a sense of camaraderie among owners.

In terms of awards and accolades, the Barracuda held its own against contemporary rivals. It frequently appeared on "Best Muscle Cars" lists in automotive publications and won its share of comparison tests. The 1970-71 Hemi 'Cuda, in particular, was often singled out for praise. Car and Driver included it in their "25 Greatest Cars of the 20th Century" list. However, it's worth noting that awards were spread across the muscle car field, with each major player receiving recognition for their strengths.

The public perception of the Barracuda evolved over its lifetime. Initially seen as a compact sporty car, it grew into a respected and feared muscle car contender. By the time the E-body models arrived, the 'Cuda had established itself as a premium choice for performance enthusiasts. Its relative rarity compared to Mustangs and Camaros added to its mystique.

In retrospect, the media and public perception of the Barracuda played a significant role in cementing its status as a muscle car icon. At the same time, it may not have dominated sales charts like the Mustang or achieved the pop culture ubiquity of the Camaro, but the 'Cuda carved out its own niche. It was perceived as a serious performer, a stylish choice, and, particularly in its Hemi guise, the apex predator of the muscle car world. This perception has only grown stronger over time, contributing to the Barracuda's legendary status among classic car enthusiasts today.

Section 10.6: Racing Rivalries

The Plymouth Barracuda's competitive spirit extended far beyond the streets and showrooms, making its mark in various forms of motorsport. On the high-banked ovals of NASCAR, the 'Cuda faced off against its pony car rivals in thrilling stock car showdowns. While not as prevalent as some of its competitors, the Barracuda's appearances in NASCAR events showcased its potential as a formidable race car, often punching above its weight against more established names.

However, it was in the world of drag racing where the Barracuda truly shone. The National Hot Rod Association (NHRA) circuits became a second home for many 'Cuda enthusiasts and factory-backed teams. The Hemi-powered Barracudas, in particular, became legends of the quarter-mile, often dominating their classes and setting records that would stand for years. The sight of a 'Cuda launching hard off the line, its front wheels lifting skyward, became an iconic image of 1970s drag racing.

The Trans-Am Series provided yet another arena for the Barracuda to prove its mettle. This road racing championship showcased the 'Cuda's handling prowess and all-around performance capabilities. Going toe-to-toe with Mustangs, Camaros, and Challengers on twisting road courses, the Barracuda held its own, earning respect from drivers and fans alike. These races not only bolstered the car's reputation but also led to valuable developments that would find their way into production models.

At the grassroots level, the Barracuda found a home in amateur racing scenes across the country. From local drag strips to regional road racing events, 'Cuda owners consistently pitted their machines against other muscle cars. These competitions often served as proving grounds for modifications and tuning techniques, with the results filtering up to influence both factory offerings and professional racing efforts.

Behind the scenes, Plymouth's commitment to racing success was evident in its factory support programs. While not as extensive as some of its competitors, Plymouth provided crucial backing to racers flying the Barracuda flag.

This support came in various forms, from specially prepared engines and parts to technical expertise and even financial support for some teams. The knowledge gained from these racing programs directly influenced the development of high-performance street models, creating a symbiotic relationship between Plymouth's racing efforts and its production car lineup.

The Barracuda's presence in these diverse racing arenas did more than just rack up wins and trophies. It cemented the car's reputation as a performance machine, capable of going head-to-head with the best its rivals could offer. From the thunderous straightaways of NASCAR to the Christmas tree lights of NHRA drag strips, the 'Cuda proved time and again that it was a force to be reckoned with. These racing rivalries not only enhanced the Barracuda's appeal to performance-minded buyers but also contributed to the rich tapestry of American motorsport history, leaving an indelible mark that continues to be celebrated by enthusiasts today.

Section 10.7: Legacy and Collectibility

The Plymouth Barracuda's journey from showroom floor to collector's garage is a tale of enduring appeal and automotive nostalgia. As we explore the legacy and collectibility of the 'Cuda alongside its contemporaries, we uncover a fascinating story of survival, value, and enduring passion.

Survivability has become a crucial factor in the collectibility of muscle cars, and the Barracuda is no exception. While production numbers varied widely across models and years, the survival rate of 'Cudas, particularly the high-performance variants, is relatively low.

Unleashing the Beast: A Comprehensive History of the Barracuda

This scarcity has contributed significantly to their desirability among collectors. For instance, the legendary 1970-1971 Hemi 'Cudas, with their limited production run, have become some of the most sought-after muscle cars in existence. In comparison, more common models like the Ford Mustang have a higher survival rate, which impacts their relative collectibility and value.

Auction results over the past few decades have painted a clear picture of the Barracuda's ascension in the collector car market. While all classic muscle cars have seen an appreciation in value, specific Barracuda models have experienced meteoric rises in value.

The aforementioned Hemi 'Cudas, especially convertibles, have fetched millions at auction, often outperforming comparable models from other manufacturers. Even more modest Barracuda variants have seen steady value increases, though they may not reach the stratospheric prices of their Hemi-powered siblings. This trend reflects not only the car's performance heritage but also its unique place in muscle car history.

Restoration challenges play a significant role in the collectibility equation. Parts availability for Barracudas can be more limited compared to some competitors, particularly for rare models or unique trim pieces. This scarcity can make restoration projects more challenging and expensive, but it also adds to the exclusivity of a fully restored example. In contrast, cars like the Mustang or Camaro often benefit from more robust aftermarket support, making restorations more accessible but potentially less exclusive.

The modern-day perception of the Barracuda among enthusiasts has undergone significant evolution. While it may not have the same widespread recognition as the Mustang or Camaro, the 'Cuda has cultivated a devoted following. Enthusiasts appreciate its unique styling, particularly of the third-generation models, and its performance pedigree.

Unleashing the Beast: A Comprehensive History of the Barracuda

The Barracuda is often seen as a connoisseur's choice, a muscle car that represents the pinnacle of the era for those in the know. This perception contrasts with the more mainstream appeal of some of its contemporaries, adding to its mystique and collector appeal.

The impact of the Barracuda on future models extends beyond its direct lineage. While Plymouth and the Barracuda name are no longer in production, the car's influence can be seen in modern muscle cars and performance vehicles.

The limited-edition, high-performance ethos of cars like the Challenger Hellcat and Demon owes much to the spirit of the Hemi 'Cuda. Moreover, the Barracuda's legacy has kept alive discussions of its potential revival, with rumors periodically surfacing about a new Barracuda model, testament to its enduring appeal.

In the grand tapestry of muscle car history, the Barracuda holds a unique position. Its rarity, performance pedigree, and distinctive styling have cemented its status as a blue-chip collector car. While it may not have outsold its rivals in its heyday, in the collector market, specific Barracuda models now command a premium that surpasses many of their period competitors.

This reversal of fortunes in the collector market underscores the Barracuda's special place in automotive history. As we reflect on the legacy and collectibility of the Barracuda, it's clear that its journey from showroom to show field has been as dramatic as its performance on the street and strip.

The 'Cuda has transcended its role as a mere automobile to become a rolling piece of American history, a testament to an era of unbridled performance and style, and a cherished icon for collectors and enthusiasts alike.

Unleashing the Beast: A Comprehensive History of the Barracuda

Chapter 11: The Barracuda's Cultural Impact and Legacy

Section 11.1: The Barracuda in Popular Media

The Plymouth Barracuda's influence extends far beyond the realm of automotive enthusiasts, permeating popular culture in ways that have cemented its status as an icon of American muscle. From the silver screen to the airwaves, the 'Cuda has left an indelible mark on our collective consciousness.

Hollywood's love affair with the Barracuda has been long-standing and passionate. The car's sleek lines and powerful presence have made it a favorite among filmmakers looking to add a touch of automotive charisma to their productions. Perhaps the most famous on-screen Barracuda is the yellow 1971 convertible driven by Don Johnson's character in the hit TV series "Nash Bridges." This particular 'Cuda became as much a star of the show as Johnson himself, showcasing the car's ability to capture attention and imagination even decades after its production ceased.

The Barracuda's influence on music is equally profound. Heart's 1977 hit song "Barracuda" may not have been directly inspired by the car, but it certainly helped reinforce the fierce and powerful image associated with the name. The driving rhythm and aggressive guitar riffs of the song embody the spirit of the muscle car era, further intertwining the Barracuda with the cultural zeitgeist of its time.

In the realm of advertising and marketing, the Barracuda left an indelible mark with clever and memorable campaigns. One of the most iconic was the "The fish that ate the competition" slogan, which played on the car's aquatic namesake to highlight its competitive edge in the muscle car market. This campaign, along with others, helped establish the Barracuda's image as a formidable contender in the pony car wars of the 1960s and early 1970s.

As technology has evolved, so too has Barracuda's presence in popular media. The digital age has seen the 'Cuda featured prominently in various racing games and driving simulators. The Forza Motorsport series, for example, has included multiple Barracuda models over the years, allowing a new generation of enthusiasts to experience the thrill of driving these legendary machines, albeit virtually. This digital preservation ensures that the Barracuda's legacy continues to reach new audiences and remains relevant in contemporary car culture.

Literature and automotive journalism have also played a crucial role in maintaining and enhancing the Barracuda's mystique. Numerous books, magazines, and articles have been devoted to examining the car's history, technical specifications, and cultural significance.

Particularly influential were the road tests and reviews from the 1960s and 1970s, which not only provided contemporary assessments of the Barracuda's performance but also helped shape public perception of the car. These period pieces now serve as valuable historical documents, offering insights into how the Barracuda was received during its heyday.

Unleashing the Beast: A Comprehensive History of the Barracuda

The Barracuda's enduring presence in popular media speaks to its timeless appeal. Whether it's tearing up the streets in a blockbuster movie, providing the namesake for a rock anthem, or being lovingly recreated in a digital environment, the Plymouth Barracuda continues to captivate audiences across various platforms. This widespread representation ensures that even those who may never have seen a Barracuda in person can appreciate its significance and allure, solidifying its place not just in automotive history but in the broader tapestry of American popular culture.

Section 11.2: The Barracuda's Influence on Car Design

The Plymouth Barracuda's impact on automotive design is undeniable. From its inception in 1964 to its final year in 1974, the Barracuda played a pivotal role in shaping the aesthetics and design language of muscle cars and beyond. This section explores the various ways in which the Barracuda left its mark on the automotive world.

One of the Barracuda's most significant contributions to car design was its pioneering of the fastback silhouette. When the first-generation Barracuda debuted in 1964, its large, wraparound rear window was a bold departure from the conventional designs of the time.

This distinctive feature not only sets the Barracuda apart from its competitors but also influenced other manufacturers to adopt similar styling cues. The Ford Mustang, for instance, introduced its own fastback variant in 1965, just a year after the Barracuda's debut. This trend towards sleek, aerodynamic profiles would continue to influence car design for decades to come.

As the Barracuda evolved, it played a crucial role in shaping muscle car aesthetics. The transition from the first generation to the second in 1967 saw a more aggressive stance and muscular proportions, setting the stage for the iconic third-generation E-body design of 1970-1974. This final iteration of the Barracuda, with its vast, low-slung body, long hood, and short deck, epitomized the classic muscle car look. Its influence can be seen in contemporaries like the Dodge Challenger and even in modern muscle car revivals.

The Barracuda wasn't just about exterior design; it also made significant contributions to automotive interiors. One of the most memorable and unique features was the "Mod Top" vinyl roof option, introduced in 1969. This psychedelic pattern, available in various colors, was a bold statement that perfectly captured the spirit of the late 1960s. While it may seem kitschy today, the Mod Top demonstrated Plymouth's willingness to push design boundaries and cater to youth culture. This approach would influence automotive marketing and design for years to come.

Color and graphics packages were another area where the Barracuda left an indelible mark. The introduction of High Impact colors in 1970 brought eye-popping hues like "Lime Light," "Vitamin C," and "In-Violet" to the streets.

These bold color choices, combined with striking graphics packages like the iconic "hockey stick" stripes, helped define the visual language of the muscle car era. The Barracuda's daring use of color and graphics inspired competitors and continues to influence special edition paint schemes and graphics on modern performance cars.

Perhaps most importantly, the Barracuda showcased how form could follow function in performance-oriented design. Features like the optional shaker hood, which protruded through the hood and moved in sync with the engine's vibrations, were not only visually striking but also served a practical purpose by enhancing engine cooling. Similarly, the rear window louvers, while aesthetically pleasing, also helped reduce interior heat and glare. These elements demonstrated that performance cars could be both beautiful and functional, a philosophy that continues to guide performance car design today.

The Barracuda's influence extended beyond its contemporaries and can still be seen in modern car design. The emphasis on aggressive stance, bold graphics, and performance-oriented aesthetics that the Barracuda helped popularize continues to inform the design of today's muscle cars and sports coupes. The recent revival of retro-inspired designs, seen in vehicles like the fifth-generation Chevrolet Camaro and the modern Dodge Challenger, owes much to the groundwork laid by cars like the Barracuda.

In conclusion, the Plymouth Barracuda's contributions to car design are far-reaching and enduring. From pioneering the fastback profile to setting trends in color and graphics, the Barracuda helped shape the visual language of performance cars. Its influence can be seen not only in its contemporaries but also in modern vehicle design, cementing its place as a true icon of automotive aesthetics. The Barracuda's legacy serves as a testament to the power of bold, innovative design in capturing the imagination of car enthusiasts and the general public alike.

Section 11.3: The Barracuda Community

The Plymouth Barracuda's impact extends far beyond its years of production, fostering a vibrant and dedicated community of enthusiasts that continues to thrive decades after the last 'Cuda rolled off the assembly line. This passionate group of fans has played a crucial role in preserving the car's legacy and keeping its spirit alive.

Car clubs and enthusiast groups have been at the heart of the Barracuda community since the vehicle's inception. Organizations like the Plymouth Owners Club have long championed the Barracuda, with dedicated subgroups focused explicitly on the 'Cuda and its variants. These clubs serve as invaluable resources for owners and admirers alike, offering technical expertise, parts sourcing advice, and a sense of camaraderie among like-minded individuals. Local chapters organize regular meetups, cruise nights, and tech sessions, allowing members to share their love for the Barracuda in person.

The digital age has revolutionized how Barracuda enthusiasts connect and share information. Online forums and social media platforms have created global communities where fans can discuss everything from restoration tips to historical trivia. Popular Facebook groups dedicated to the 'Cuda boast thousands of members who regularly share photos, stories, and advice. These online spaces have become virtual garages where enthusiasts can troubleshoot problems, showcase their projects, and connect with fellow Barracuda lovers from around the world.

Car shows and meetups remain essential events for the Barracuda community. The Chrysler Nationals at Carlisle, Pennsylvania, has become a mecca for Mopar enthusiasts, with a strong showing of Barracudas each year.

These events not only allow owners to display their prized possessions but also serve as networking opportunities and marketplaces for rare parts and memorabilia. The sense of excitement and nostalgia at these gatherings is palpable, as attendees marvel at pristine restorations and rare factory options.

The preservation and restoration of Barracudas have given rise to a robust network of businesses and resources supporting the community. Specialty parts manufacturers produce everything from body panels to interior trim, enabling enthusiasts to bring even the most neglected 'Cudas back to life. Detailed restoration guides, both in print and online, provide step-by-step instructions for tackling complex projects. This ecosystem of support ensures that the knowledge and materials needed to keep Barracudas on the road remain accessible to future generations.

Perhaps most encouraging for the Barracuda's long-term legacy is the continued engagement of younger enthusiasts. While many classic car communities struggle with an aging demographic, the Barracuda has managed to captivate new generations of fans. Social media influencers and young car builders showcasing Barracuda projects on platforms like YouTube and Instagram have introduced the 'Cuda to audiences who may have never seen one in person. This influx of young blood ensures that the Barracuda's story will continue to be told and its influence felt for years to come.

The Barracuda community is more than just a group of car enthusiasts; it's a testament to the enduring appeal of a true American icon. Through their passion, dedication, and shared knowledge, these enthusiasts have transformed the Barracuda from a discontinued model into a living legend. As long as this community continues to thrive, the spirit of the Plymouth Barracuda will roar on, inspiring new generations of automotive enthusiasts and ensuring its place in the pantheon of classic American muscle cars.

Section 11.4: The Barracuda's Impact on the Collector Car Market

The Plymouth Barracuda's influence extends far beyond its brief production run, leaving an indelible mark on the collector car market. As time has passed, the Barracuda has transformed from a popular muscle car into a highly sought-after collector's item, with specific models achieving legendary status among enthusiasts and investors alike.

At the heart of the Barracuda's collector appeal lies its rarity. While production numbers were substantial during its heyday, the passage of time and the car's popularity have significantly reduced the number of surviving examples, particularly those in pristine condition. This scarcity has dramatically affected the Barracuda's collectibility, with specific models becoming automotive holy grails.

The most coveted among these are undoubtedly the HEMI 'Cudas, especially the convertible variants. These rare beasts have achieved astronomical values at auctions, with some examples fetching millions of dollars. For instance, a 1971 HEMI 'Cuda convertible, one of only 11 ever produced, sold for a staggering $3.5 million at a Mecum auction in 2014, solidifying its status as one of the most valuable muscle cars ever sold.

The Barracuda's journey from performance car to prized collectible has positioned it as a legitimate financial asset in the classic car world. Savvy investors and enthusiasts alike have recognized the potential for significant returns on well-maintained or restored Barracudas. Price trends over the past few decades have shown a consistent upward trajectory, with even non-HEMI models appreciating considerably. For example, a well-preserved 1970 Barracuda Gran Coupe that might have sold for $15,000 in the late 1990s could easily command over $50,000 today, with pristine examples fetching even more.

This rise in value has given birth to a thriving restoration economy centered around the Barracuda. The demand for authentic parts and expert restoration services has created a robust aftermarket industry. Businesses specializing in E-body parts and restoration services have flourished, offering everything from small trim pieces to complete body shells for ground-up restorations. This economic impact extends beyond just parts, encompassing skilled labor, specialty tools, and even niche insurance products tailored for high-value muscle cars.

The auction world has played a significant role in elevating the Barracuda's status and value. Record-breaking sales at prestigious auctions have not only set new price benchmarks but have also thrust the Barracuda into the spotlight of mainstream media. These high-profile sales have garnered attention from beyond the traditional car collector community, attracting new enthusiasts and investors. The aforementioned 1971 HEMI 'Cuda convertible sale is just one example of how these auctions have cemented the Barracuda's place in the upper echelons of collector cars.

However, with great value comes great temptation. The high prices commanded by rare Barracudas have led to a rise in counterfeits and clones. Less scrupulous sellers may attempt to pass off a standard Barracuda as a more valuable model, or even create a complete replica of a rare variant. This has created a challenge for authenticating genuine examples, particularly when it comes to the most valuable models like HEMI 'Cudas. Serious collectors and auction houses have developed rigorous authentication processes, often involving marque experts and extensive documentation to verify a car's provenance. Methods for ascertaining genuine HEMI 'Cudas include checking original broadcast sheets, decoding VINs, and scrutinizing specific components known to be unique to these rare models.

The Barracuda's impact on the collector car market goes beyond mere dollars and cents. It represents a tangible link to a bygone era of American automotive prowess, a time when raw power and bold styling reigned supreme. For many collectors, owning a Barracuda is about more than investment potential; it's about preserving a piece of automotive history and experiencing the thrill of a true muscle car icon. As the collector car market continues to evolve, the Plymouth Barracuda stands as a testament to the enduring appeal of classic American muscle. Its rarity, performance pedigree, and cultural significance ensure that it will remain a highly coveted prize for collectors and enthusiasts for generations to come.

Whether as a sound investment, a restoration project, or simply a dream car to be admired, the Barracuda's place in the pantheon of collectible automobiles is secure, its value extending far beyond its original sticker price to become a true automotive legend.

Section 11.5: The Barracuda's Technological Legacy

The Plymouth Barracuda wasn't just a pretty face in the muscle car era; it was a technological powerhouse that left an indelible mark on automotive engineering. This section explores the lasting impact of the Barracuda's innovations on the automotive industry.

The heart of any muscle car is its engine, and the Barracuda's powerplants were truly revolutionary. The crown jewel was undoubtedly the 426 HEMI, a behemoth that continues to influence modern MOPAR engines. This legendary motor, with its hemispherical combustion chambers, set new standards for power output and efficiency. Today, we see its legacy in the supercharged HEMI engines powering modern Dodge muscle cars. The Barracuda's role in popularizing high-performance V8 engines for street use cannot be overstated, paving the way for the horsepower wars that continue to this day.

But the Barracuda's innovations weren't limited to raw power. Its suspension and handling advancements were equally impressive. The E-body Barracuda's wide stance and refined suspension geometry offered a level of handling prowess that was uncommon for American muscle cars of the era. This approach to balancing straight-line speed with cornering ability influenced future performance car designs, challenging the notion that muscle cars were only good for quarter-mile runs.

Safety features, often overlooked in discussions of muscle cars, were another area where the Barracuda made significant contributions. As safety regulations became more stringent in the late 1960s and early 1970s, the Barracuda incorporated innovative features, such as collapsible steering columns and improved braking systems. These advancements not only made the Barracuda safer but also influenced future safety regulations and designs across the industry.

The manufacturing techniques used to produce the Barracuda also left a lasting impact. The E-body's unibody construction was a leap forward in creating a rigid yet relatively lightweight platform for high-performance applications. This approach to chassis design would become increasingly common in the decades that followed, as automakers sought to improve both performance and efficiency.

One of the most enduring legacies of the Barracuda is its influence on the performance aftermarket. The Barracuda became a favorite canvas for hot rodders and performance enthusiasts, spurring the development of a vast array of aftermarket parts and accessories. From high-flow cylinder heads to advanced suspension components, many iconic aftermarket parts made their debut on the Barracuda.

This symbiotic relationship between factory muscle cars and the aftermarket continues to thrive today, with modern muscle cars benefiting from the lessons learned during the Barracuda era.

The Barracuda's technological innovations weren't just about raw numbers or spec sheets. They represented a philosophy of continuous improvement and pushing boundaries. Engineers and designers working on the Barracuda were constantly seeking ways to make the car faster, safer, and more exciting to drive. This spirit of innovation continues to inspire automotive engineers today.

It's worth noting that many of Barracuda's advancements occurred during a period of rapid change in the automotive industry. Stricter emissions regulations and safety standards were forcing manufacturers to rethink their approach to performance. The Barracuda's ability to adapt and innovate in this challenging environment is a testament to the ingenuity of its creators. Today, we can see the Barracuda's technological DNA in many modern performance cars.

The balance of power, handling, and style that the Barracuda exemplified remains a gold standard. From advanced engine management systems that can trace their roots back to the fine-tuning required for high-performance carburetors, to suspension designs that still follow principles established during the muscle car era, the Barracuda's influence is far-reaching.

In conclusion, the Plymouth Barracuda's technological legacy extends far beyond its production years. Its innovations in engine design, chassis engineering, safety features, and manufacturing techniques continue to influence the automotive industry. The Barracuda served as a proving ground for ideas that would shape the future of performance cars, and its impact can still be felt in the roar of modern engines and the precision of contemporary handling dynamics. As we look to the future of automotive technology, we do so standing on the shoulders of giants like the Plymouth Barracuda.

Section 11.6: The Barracuda's Environmental and Social Context

The Plymouth Barracuda, like many of its muscle car contemporaries, was born into a rapidly changing world. Its rise and fall were inextricably linked to the social, economic, and environmental shifts of the 1960s and early 1970s. This section examines the external factors that influenced Barracuda's journey and ultimately contributed to its demise.

The 1973 Oil Crisis dealt a severe blow to the muscle car era, and the Barracuda was no exception. As fuel prices skyrocketed and shortages led to long lines at gas stations, American consumers suddenly found themselves prioritizing fuel efficiency over raw power. The Barracuda, with its thirsty V8 engines, quickly fell out of favor. Chrysler scrambled to adapt, but the writing was on the wall for the gas-guzzling pony car.

The shift in priorities wasn't limited to fuel economy. The early 1970s saw a rapid implementation of stringent emissions regulations, presenting a significant challenge for high-performance vehicles like the Barracuda. Engineers struggled to maintain the car's power output while meeting new environmental standards. This battle was evident in the changes to the Barracuda's engine offerings between 1970 and 1974. The mighty HEMI and high-output 440 engines were phased out, replaced by detuned versions that sacrificed performance in the name of cleaner exhaust.

Safety concerns also played a crucial role in the changing automotive landscape. Ralph Nader's influential book "Unsafe at Any Speed," while not specifically targeting the Barracuda, sparked a national debate about car safety. This led to new regulations that affected vehicle design across the board. The Barracuda, designed primarily for performance, had to evolve to meet these new standards. This often meant added weight and complexity, further challenging its performance credentials.

Consumer tastes were also shifting dramatically. As the baby boomer generation matured, many began to prefer personal luxury coupes over the raw sportiness of pony cars. Models like the Chevrolet Monte Carlo and Ford Thunderbird gained popularity, offering a more comfortable and refined driving experience. The Barracuda, despite attempts to provide luxury options, was unable to shake its performance-oriented image and capture this emerging market.

Despite these challenges, the Barracuda remained a potent symbol of American automotive prowess. It stood toe-to-toe with European sports cars of the era, showcasing what American engineering and design could achieve. The Barracuda's bold styling, powerful engines, and racing success on both drag strips and road courses demonstrated that American cars could compete on a global stage.

The Barracuda's journey through this tumultuous period reflects the broader story of American society in the early 1970s. It represents the end of an era characterized by unbridled optimism and the beginning of a more cautious, environmentally conscious age. The tensions between performance, safety, and environmental responsibility that the Barracuda faced continue to shape the automotive industry today.

In many ways, the Barracuda was a casualty of its time, a victim of forces beyond its control. Yet, its struggle to adapt to these changes and its ultimate discontinuation have only added to its mystique. The Barracuda stands as a time capsule of sorts, encapsulating the spirit of an era when performance reigned supreme and the open road beckoned with the promise of adventure and freedom.

Section 11.7: The Barracuda's Enduring Mystique

The Plymouth Barracuda, despite being out of production for nearly five decades, continues to captivate automotive enthusiasts and casual observers alike. This enduring mystique is a testament to the car's iconic status and the indelible mark it left on American car culture.

At the heart of the Barracuda's lasting appeal is the powerful nostalgia it evokes. For many Baby Boomers, the Barracuda represents more than just a car; it's a time machine that transports them back to their youth. The sight, sound, and even smell of a well-preserved 'Cuda can trigger a flood of memories, cruising down Main Street on a warm summer night, the thrill of a first date, or the pride of owning one's first high-performance vehicle. This emotional connection has driven many to seek out and restore Barracudas, not just as collector's items, but as tangible links to their past.

The Barracuda's mystique is further enhanced by the tantalizing "what if" factor. Its abrupt discontinuation in 1974 left fans wondering how the model might have evolved had it survived the tumultuous 1970s. This speculation has given rise to a vibrant subculture of automotive enthusiasts who create stunning renderings and even full-scale concept cars, imagining modern interpretations of the Barracuda.

These "fantasy Barracudas" often feature sleek, updated designs that pay homage to the original while incorporating contemporary performance and technology. The popularity of these concepts underscores the continued relevance and appeal of the Barracuda's design ethos.

The influence of the Barracuda on modern muscle cars is undeniable. When the Big Three American automakers revived their muscle car lineups in the 21st century, the spirit of the Barracuda was evident. While Dodge didn't directly resurrect the Barracuda name, the modern Challenger, particularly in its high-performance variants, clearly channels the aggressive stance and powerful presence of its E-body predecessor. The Challenger's success in the contemporary market is a testament to the enduring appeal of the design philosophy pioneered by cars like the Barracuda.

Unleashing the Beast: A Comprehensive History of the Barracuda

In recent years, the Barracuda has become a favorite subject for resto-mod projects. This trend involves taking the classic Barracuda body and enhancing it with modern performance components, electronics, and comfort features. High-profile builds showcased at events like the SEMA Show have demonstrated how the Barracuda's timeless design can be seamlessly merged with cutting-edge automotive technology. These resto-mods represent the best of both worlds, classic style with modern performance, and have introduced the Barracuda to a new generation of enthusiasts.

Ultimately, the Barracuda's place in automotive history is secure. Its inclusion in prestigious automotive museums and historical collections around the world confirms its status as a true icon of the muscle car era. The Barracuda represents more than just high performance or striking design; it embodies a pivotal moment in American cultural history, when youth, freedom, and horsepower converged to create something truly special.

The enduring mystique of the Plymouth Barracuda is a phenomenon that transcends mere automotive enthusiasm. It speaks to our collective nostalgia for a bygone era, our appreciation for bold design and engineering, and our continued fascination with speed and power. As long as there are winding roads and the spirit of adventure, the legend of the Barracuda will continue to inspire and excite automotive enthusiasts for generations to come.

Unleashing the Beast: A Comprehensive History of the Barracuda

Unleashing the Beast: A Comprehensive History of the Barracuda

Chapter 12: Collecting and Restoring: The Barracuda Today

Section 12.1: The Barracuda Collector's Market

The Plymouth Barracuda, once a fierce competitor in the pony car wars of the 1960s and early 1970s, has evolved into a highly sought-after collectible in the classic car market. Over the past decade, the Barracuda has experienced a significant surge in both popularity and value, cementing its status as one of the most desirable muscle cars of its era.

Current market trends for Plymouth Barracudas have been overwhelmingly positive, with prices steadily climbing across all models and years. This upward trajectory is particularly pronounced for the rarest and most powerful variants. For instance, a well-preserved 1970 Hemi 'Cuda convertible sold for a record-breaking $3.5 million at auction in 2014, showcasing the incredible heights that top-tier Barracudas can reach in terms of value.

Among the most sought-after Barracuda models are the 1970-1971 Hemi 'Cuda variants, especially the convertibles. These represent the pinnacle of Barracuda performance and rarity, commanding prices that place them among the most valuable muscle cars in the world. However, it's not just the Hemi-powered models that attract attention. The 440-equipped vehicles, as well as those with the 340 and 383 engines, also enjoy strong demand from collectors.

Several factors affect Barracuda's values in today's market. Rarity is a primary consideration, with limited production numbers driving up prices for specific models and option combinations. Originality is another crucial factor; original, numbers-matching Barracudas with documented history can command significantly higher prices than restored examples.

The level of documentation, including original build sheets, window stickers, and ownership history, can substantially impact a car's value. Condition, of course, plays a vital role, with pristine, low-mileage examples fetching premium prices.

For those looking to enter the Barracuda market or add to their collection, there are various avenues to explore. Traditional classic car auctions, both live and online, remain popular sources for high-end Barracudas. However, websites like Hemmings and BringATrailer have become increasingly popular platforms for buying and selling classic Barracudas of all conditions and price points. Local classic car dealerships, enthusiast forums, and car clubs can also be excellent resources for finding Barracudas for sale.

When considering the investment potential of Barracudas, it's essential to approach with both enthusiasm and caution. While not all Barracudas have seen astronomical price increases, specific rare models have consistently outperformed traditional investments. The most desirable variants, particularly those with documented histories and in excellent condition, have shown strong appreciation over time. However, as with any investment, there are risks to consider. Market fluctuations, changes in collector preferences, and the overall economy can all impact classic car values.

For those passionate about these iconic pony cars, collecting Barracudas offers more than just potential financial rewards. It provides an opportunity to own and preserve a piece of automotive history, to experience the raw power and style of the muscle car era, and to join a vibrant community of fellow enthusiasts. Whether you're drawn to the sleek lines of a first-generation Barracuda or the aggressive stance of a '71 Cuda, there's a place for you in the world of Barracuda collecting.

As the classic car market continues to evolve, the Plymouth Barracuda remains a shining star, its allure undiminished by the passage of time. For collectors and enthusiasts alike, the Barracuda represents not just a car, but a tangible connection to one of the most exciting periods in American automotive history.

Section 12.2: Restoration Challenges and Considerations

Restoring a Plymouth Barracuda is a labor of love that comes with its own unique set of challenges and considerations. Before embarking on a restoration project, it's crucial to assess the condition of your Barracuda thoroughly. A comprehensive inspection of the body, frame, and mechanical components will provide a clear picture of the work ahead and help you plan your restoration strategy.

One of the primary challenges in Barracuda restoration is addressing rust and structural issues. The Barracuda's unibody construction, while innovative for its time, can make structural repairs more complex than those on body-on-frame vehicles. Rust is a common enemy, particularly in the rear quarter panels, floor pans, and trunk areas. Skilled metalwork is often required to adequately address these issues and maintain the car's structural integrity.

Parts sourcing can be another significant hurdle in restoring a Barracuda. While many components are shared with other Mopar vehicles of the era, some Barracuda-specific parts can be challenging to find. This is especially true for rare models or trim levels. Restorers often find themselves scouring swap meets, online marketplaces, and specialty suppliers to locate the correct parts. In some cases, fabrication of custom parts may be necessary when original components are unavailable.

The debate between maintaining originality and incorporating modifications is a constant in the Barracuda restoration community. While some purists insist on keeping everything stock, adhering to factory specifications and using period-correct parts, others opt for a resto-mod approach. Resto-mods enable the addition of modern conveniences and performance upgrades while preserving the classic aesthetic of the Barracuda. This can include updated engines, modern brake systems, improved suspension, and subtle interior upgrades, such as hidden audio systems or air conditioning.

Cost is a significant consideration in any restoration project, and Barracuda restorations can be costly, especially for rare or high-performance models. A complete, show-quality restoration of a Hemi 'Cuda can easily exceed $200,000, not including the initial purchase price of the car. Even more modest restorations can run into tens of thousands of dollars. It's essential to have a realistic budget and to factor in unexpected expenses that invariably arise during the restoration process.

Time is another crucial factor to consider. A complete, frame-off restoration can take anywhere from 18 months to several years, depending on the car's condition, the extent of the restoration, and the restorer's resources and expertise. Several factors can impact the timeline, including the availability of parts, the complexity of the repairs required, and whether the work is being performed by a professional shop or as a DIY project.

Documentation and research play a vital role in a successful Barracuda restoration. Obtaining original build sheets, factory literature, and period photographs can be invaluable in ensuring accuracy during the restoration process. This is particularly important for rare or high-value Barracudas, where historical accuracy can significantly impact the car's value.

Lastly, it's essential to consider the end goal of your restoration. Are you aiming for a concours-quality show car, a faithful driver-quality restoration, or a personalized resto-mod? Each path comes with its own set of challenges and considerations, from sourcing date-coded original parts for a concours restoration to selecting appropriate modern upgrades for a resto-mod.

Restoring a Plymouth Barracuda is a rewarding but challenging endeavor. It requires patience, dedication, and often a significant investment of time and resources. However, for those passionate about preserving these iconic muscle cars, the result is a beautifully restored piece of automotive history well worth the effort.

Section 12.3: Restoration Resources and Expertise

The journey of restoring a Plymouth Barracuda can be both challenging and rewarding. Fortunately, a wealth of resources and expertise is available to help enthusiasts bring these iconic pony cars back to their former glory. This section explores the various avenues for finding parts, accessing specialized knowledge, and connecting with the broader Barracuda community.

One of the primary concerns for any restoration project is sourcing the necessary parts. For Barracuda restorers, several companies specialize in reproduction parts that cater specifically to these classic Mopars. Companies like Year One and Classic Industries offer extensive catalogs of reproduction parts for Barracudas, ranging from body panels and trim pieces to interior components and mechanical parts. These reproduction parts are often manufactured to exacting standards, closely mimicking the original equipment in both form and function.

For those seeking original parts, the hunt can be more challenging but potentially more rewarding. Swap meets, online marketplaces, and specialty salvage yards can be treasure troves for hard-to-find original components. The Mopar community is known for its dedication to preserving these classics, and many enthusiasts maintain extensive inventories of original parts.

When it comes to the actual restoration work, some Barracuda owners choose to entrust their prized possessions to professional restoration specialists. Shops like Legendary Motorcar Company have built reputations for high-quality Barracuda restorations. These specialists often have years of experience working specifically with Mopars and can handle everything from minor repairs to complete frame-off restorations. While professional restorations can be costly, they frequently result in show-quality vehicles that can compete at the highest levels of judged events.

For the do-it-yourself restorer, a wealth of resources is available to guide you through the process. Detailed restoration manuals specific to the Barracuda can provide step-by-step instructions for various aspects of the restoration process. Online forums like "For A Bodies Only" have become invaluable resources for Barracuda owners tackling their own restorations. These communities allow enthusiasts to share knowledge, troubleshoot problems, and seek advice from those who have already completed similar projects.

Video tutorials and online courses have also become increasingly popular, offering visual guides to various restoration techniques. Many experienced restorers and mechanics have created YouTube channels dedicated to Mopar restorations, providing a wealth of free information to the DIY community.

Documentation plays a crucial role in any restoration project, both in terms of the car's history and the restoration process itself. The Chrysler Historical Services can provide build sheets and other valuable documentation for your Barracuda, offering insights into its original configuration and options. This information is invaluable for ensuring historical accuracy in your restoration.

Networking within the Barracuda community can open doors to a wealth of knowledge and resources. Joining clubs like the Plymouth

Owners Club not only provides access to a network of fellow enthusiasts but often includes benefits such as club magazines, technical support, and access to members-only events. These connections can be invaluable when seeking advice, tracking down rare parts, or simply sharing your passion for these remarkable vehicles.

Local car shows and cruise nights can also be excellent opportunities to connect with other Barracuda owners and enthusiasts in your area. These events often attract a wealth of knowledge and experience, and many attendees are more than happy to share their insights and experiences with fellow enthusiasts.

As you embark on your Barracuda restoration journey, remember that patience and perseverance are key. Restoration projects often take longer and cost more than initially anticipated, but the result, a beautifully restored piece of American automotive history, is well worth the effort. By tapping into the wealth of resources and expertise available, you'll be well-equipped to tackle the challenges and enjoy the rewards of bringing a classic Barracuda back to life.

Section 12.4: Preserving Originality and Authenticity

In the world of classic car restoration, preserving originality and authenticity is paramount, especially when it comes to iconic vehicles like the Plymouth Barracuda. For many collectors and enthusiasts, maintaining the car's original specifications and components is not just about preserving history; it's about maximizing value and enjoying a genuine piece of automotive heritage.

One of the most critical aspects of preserving a Barracuda's authenticity is maintaining its numbers-matching components. A numbers-matching Barracuda, particularly a Hemi 'Cuda, can be worth significantly more than one with a non-original engine or drivetrain. This premium is due to the rarity and historical significance of having the original factory-installed components still in place. When restoring a Barracuda, great care should be taken to preserve and refurbish these original parts whenever possible, rather than replacing them with newer components.

Equally important in the quest for authenticity are the correct finishes and materials used throughout the car. Using period-correct paints, materials, and finishes in restoration is crucial for maintaining the Barracuda's original appearance and feel. This attention to detail extends to every aspect of the car, from the exact shade of High Impact paint like Tor-Red or Lemon Twist, to the correct grain of vinyl on the seats, and even the proper plating on nuts and bolts. For serious collectors and concours-level restorations, even the most minor details matter.

Factory options and configurations play a significant role in a Barracuda's authenticity and value. Maintaining the original option combination, such as a Shaker hood on a '71 'Cuda, is key to preserving its historical accuracy. This extends to interior options, performance packages, and even dealer-installed accessories. Deviating from the car's original configuration can significantly impact its value and appeal to purist collectors.

When it comes to sourcing parts for a restoration, collectors often face the choice between New Old Stock (NOS) parts and modern reproductions. NOS parts, which are original, unused components from the period, offer the ultimate in authenticity. However, they can be scarce and prohibitively expensive. Quality reproductions, on the other hand, can provide a more cost-effective alternative for many components while still maintaining a high degree of authenticity. The key is to use NOS parts in visible or critical areas where originality is most important, and high-quality reproductions where appropriate.

Throughout the restoration process, documenting every step is crucial. Keeping detailed records and photographs of your restoration can significantly enhance your Barracuda's value and provenance. This documentation serves multiple purposes: it provides proof of the work done, helps maintain the car's history, and can be invaluable if the vehicle is ever sold or shown in competitions. Photos of the car before, during, and after restoration, along with receipts for parts and services, as well as any historical documentation about the car's origins, all contribute to its overall value and appeal to future collectors.

Preserving originality and authenticity in a Barracuda restoration is a delicate balance of respecting history, maintaining value, and creating a vehicle that can be enjoyed and appreciated. While it may be tempting to make upgrades or modifications, staying true to the car's original specifications often yields the most significant rewards, both in terms of monetary value and historical significance. For many Barracuda enthusiasts, the thrill of owning and driving a piece of automotive history in its original form is the ultimate goal of collecting and restoring it.

Section 12.5: The Resto-Mod Movement

The resto-mod movement has gained significant traction in the Barracuda community, offering a compelling blend of classic styling and modern performance. Resto-mods, short for restored and modified, represent a unique approach to classic car ownership that combines the timeless aesthetics of vintage Barracudas with cutting-edge technology and performance upgrades.

At its core, the resto-mod concept allows enthusiasts to enjoy the best of both worlds. These builds maintain the iconic silhouette and character of the original Barracuda while incorporating modern amenities and performance enhancements that dramatically improve the driving experience. The result is a classic muscle car that can hold its own against contemporary performance vehicles while turning heads with its vintage charm.

One of the most popular aspects of resto-mod Barracudas is the integration of modern powertrains. Many builders opt for current-generation Hemi engines, such as the 6.4-liter (392 cubic inch) or even the supercharged 6.2-liter Hellcat motor. These modern power plants offer reliability, fuel efficiency, and performance that far surpass the original engines. For instance, a resto-mod 'Cuda with a Hellcat engine can easily produce over 700 horsepower, more than doubling the output of even the most potent factory offerings from the muscle car era.

Performance upgrades extend beyond the engine bay. Resto-mod Barracudas often feature modern suspension systems, including coilover setups and upgraded sway bars, which vastly improve handling and ride quality. Brake systems are typically upgraded to modern disc brakes all around, often with large, multi-piston calipers that provide stopping power to match the enhanced performance.

These improvements transform the driving dynamics of the Barracuda, making it more suited to contemporary driving conditions and expectations. Comfort and convenience modifications are another hallmark of the resto-mod movement. Air conditioning systems designed to integrate into the classic interior seamlessly provide much-needed comfort during summer cruises.

Modern audio systems, often hidden behind stock-appearing faceplates or tucked away entirely, offer high-fidelity sound and connectivity options like Bluetooth and satellite radio. Power windows, central locking, and even keyless ignition systems can be discreetly added to enhance the user experience without compromising the classic aesthetic. One of the most challenging aspects of creating a successful resto-mod Barracuda is striking a balance between modern upgrades and classic aesthetics. The goal is to enhance the car's performance and usability without losing the essence of what makes a Barracuda special.

Skilled builders pay meticulous attention to detail, ensuring that modern additions complement rather than detract from the original design. For example, custom-built dashboards might incorporate modern gauges that mimic the style of the originals, or seats might be reupholstered with period-correct patterns but use modern materials for improved comfort and durability.

The resto-mod movement has also spawned a thriving aftermarket industry catering specifically to classic Mopar enthusiasts. Companies now offer everything from bolt-in modern chassis systems to complete bodies crafted from modern materials. This availability of parts and expertise has made it easier than ever for Barracuda owners to create their ideal blend of classic style and modern performance.

Critics of the resto-mod trend argue that these modifications detract from the historical significance and authenticity of classic Barracudas. However, proponents counter that these builds often breathe new life into cars that might otherwise be left to deteriorate. Moreover, the resto-mod approach can make classic car ownership more appealing to a younger generation of enthusiasts who appreciate vintage styling but desire modern performance and reliability.

As the resto-mod movement continues to evolve, we're seeing increasingly sophisticated builds that push the boundaries of what's possible with a classic Barracuda. From subtle upgrades that maintain a largely stock appearance to radical transformations that reimagine the Barracuda for the 21st century, the resto-mod scene offers something for every taste and preference.

Ultimately, the resto-mod movement represents a dynamic and exciting chapter in the ongoing story of the Plymouth Barracuda. It demonstrates that even decades after the last Barracuda rolled off the assembly line, enthusiasts are still finding new ways to celebrate and reimagine this iconic pony car. Whether preserving originality or embracing modern innovations, the passion for the Barracuda remains as strong as ever, ensuring its legacy will continue to thrive in the years to come.

Section 12.6: Showing and Enjoying Your Barracuda

The joy of owning a Plymouth Barracuda extends far beyond the garage. For many enthusiasts, showcasing their prized possession at car shows and participating in driving events is the ultimate reward for their dedication and hard work. The car show circuit offers Barracuda owners numerous opportunities to display their vehicles and connect with fellow enthusiasts.

One of the largest and most prestigious events for Barracuda owners is the Carlisle Chrysler Nationals, held annually in Carlisle, Pennsylvania. This event draws thousands of Mopar enthusiasts from across the country, featuring an impressive array of Barracudas alongside other classic Chrysler, Plymouth, and Dodge vehicles. It's not uncommon to see rows of gleaming 'Cudas, from rare Hemi-powered convertibles to lovingly restored base models, each with its own unique story.

When it comes to showing your Barracuda, understanding the judging criteria and classes is crucial. Many shows, such as those sanctioned by the Antique Automobile Club of America (AACA), have specific guidelines for muscle cars like the Barracuda. These criteria often focus on historical accuracy and the quality of restoration. Judges meticulously examine every aspect of the vehicle, from the accuracy of the paint color to the authenticity of small details, such as hose clamps and decals.

Preparing a Barracuda for a show competition is an art in itself. Detailing for a concours-level show often involves hours of meticulous cleaning, with enthusiasts using tools as fine as cotton swabs to clean hard-to-reach areas. Every surface, from the engine bay to the undercarriage, must be spotless. Many owners spend weeks preparing their cars, ensuring that every chrome piece gleams and every rubber seal is properly conditioned.

While static displays are popular, many Barracuda owners prefer to enjoy their cars on the open road. Driving events and rallies provide the perfect opportunity to experience these muscle cars as they were intended. The Hot Rod Power Tour, for instance, gives Barracuda owners the chance to drive their classics on a multi-day, cross-country journey. This event not only showcases the Barracuda's performance capabilities but also its ability to serve as a comfortable long-distance cruiser.

Regional clubs often organize local driving events, from scenic cruises to drag racing nights at local tracks. These events allow Barracuda owners to exercise their cars' muscles in a controlled environment, reliving the glory days of muscle car performance.

However, the classic car hobby often grapples with the debate between preservation and enjoyment. Some owners of rare, low-mileage Barracudas opt to store their prized possessions in climate-controlled facilities, preserving them as time capsules of automotive history. These cars often emerge only for the most prestigious shows or for occasional short drives to keep the mechanicals in working order.

On the other hand, many enthusiasts firmly believe that these cars were meant to be driven and enjoyed on the open road. They argue that the real value of owning a Barracuda lies in the experience of driving it, hearing the rumble of the V8 engine, and feeling the connection to automotive history with every mile.

Ultimately, the decision of how to enjoy a Barracuda comes down to personal preference. Whether it's meticulously preparing for top honors at a national show, embarking on a cross-country road trip, or simply cruising local backroads on a sunny Sunday afternoon, the Plymouth Barracuda continues to bring joy to its owners and admirers alike.

As the classic car hobby evolves, Barracuda owners are finding new ways to share their passion. Social media platforms and online forums have created virtual car shows, allowing enthusiasts to showcase their cars to a global audience. These digital platforms have also made it easier for Barracuda owners to connect, share restoration tips, and organize meetups.

Ultimately, showcasing and enjoying a Plymouth Barracuda is about more than just the car itself. It's about preserving a piece of automotive history, sharing stories, and being part of a passionate community. Whether on the show field or the open road, the Barracuda continues to turn heads and ignite enthusiasm, just as it did when it first rolled off the assembly line decades ago.

Section 12.7: The Future of Barracuda Collecting

As we look ahead, the future of Barracuda collecting promises to be as exciting and dynamic as the car itself. The landscape of classic car collecting is constantly evolving, and the Barracuda market is no exception.

In recent years, we've seen a growing interest in the often-overlooked first-generation Barracudas, as the later models become increasingly expensive and harder to find. This shift demonstrates the cyclical nature of collecting trends and suggests that even the more modest Barracuda variants may have their moment in the spotlight.

The broader automotive industry's shift towards electric vehicles is also beginning to influence the classic car world. Some companies now offer electric drivetrain conversions for classic cars, potentially extending the lifespan of Barracudas in a changing automotive landscape. While purists may balk at the idea of an electric 'Cuda, these conversions could allow enthusiasts to enjoy their beloved classics in an increasingly eco-conscious world.

Demographic shifts are playing a significant role in shaping the future of Barracuda collecting. As younger collectors enter the market, there's increased interest in the more affordable Barracuda models from the early '70s. These newer enthusiasts often bring fresh perspectives and different priorities, sometimes favoring drivability and personalisation over strict originality. This trend could lead to a resurgence in popularity for resto-mod Barracudas, blending classic style with modern performance and comfort.

Unleashing the Beast: A Comprehensive History of the Barracuda

The possibility of a modern Barracuda revival has been a topic of speculation among enthusiasts for years. While Chrysler has teased the idea of a contemporary Barracuda several times, no concrete plans have materialized as of 2023. The potential for a new Barracuda model remains an intriguing prospect that could significantly impact the collector market for original Barracudas, either by renewing interest in the classics or potentially diluting their unique appeal.

Looking at the long-term outlook for Barracuda values and collectibility, it's clear that while the muscle car market may see fluctuations, the rarest and most desirable Barracuda models are likely to remain highly sought after by collectors. The limited production numbers of specific variants, particularly the high-performance models and convertibles, ensure their continued scarcity and desirability.

However, the future of Barracuda collecting isn't just about values and investment potential. The passion for these cars runs deep, rooted in nostalgia, appreciation for automotive history, and the pure joy of experiencing a true American muscle car. As long as enthusiasts appreciate the Barracuda's unique blend of style and performance, these cars will continue to be cherished and preserved.

The Barracuda community itself will play a crucial role in shaping the future of collecting. Online forums, social media groups, and car clubs continue to connect enthusiasts from around the world, sharing knowledge, parts sources, and restoration tips. This network of support ensures that even as the cars age, the expertise needed to keep them on the road will be preserved and passed on to future generations of Barracuda fans.

Ultimately, the future of Barracuda collecting looks bright. Whether it's a meticulously preserved Hemi 'Cuda commanding seven figures at auction or a loved and driven base model Barracuda bringing joy to its owner on weekend cruises, these cars will continue to captivate automotive enthusiasts. The Plymouth Barracuda's place in automotive history is secure, and its legacy as one of the most

Unleashing the Beast: A Comprehensive History of the Barracuda

iconic muscle cars ever produced ensures its continued relevance in the collecting world for years to come.

ABOUT THE AUTHOR

Todd Bandel is an accomplished author specializing in informational history books, with a particular focus on the automotive industry. Drawing from 40 years of experience as an automotive technician, Todd combines deep expertise and passion to enlighten readers about the historical nuances of automobiles. Todd currently resides in San Diego, California, where he continues to explore and write about his enduring interest in automotive history.

Mechanicaddicts.com

www.ingramcontent.com/pod-product-compliance
Lightning Source LLC
Chambersburg PA
CBHW020657220526
45464CB00001B/469